The
CHAPANIS CHRONICLES

50 Years
of Human Factors
Research, Education,
and Design

ALPHONSE CHAPANIS

The

CHAPANIS CHRONICLES

50 Years
of Human Factors
Research, Education,
and Design

ALPHONSE CHAPANIS

Aegean Publishing Company
Santa Barbara

Published in the U.S.A. by
Aegean Publishing Company
Post Office Box 6790, Santa Barbara, California 93160

THE CHAPANIS CHRONICLES:
50 YEARS OF HUMAN FACTORS
RESEARCH, EDUCATION, AND DESIGN

ISBN 0-9636178-9-3 (hc)

If not available from your local bookstore,
this book may be ordered directly from the publisher.
Send the cover price ($34.00) plus $4.50 for shipping and handling
costs to the above address. California residents add applicable sales
tax.

Publishers Cataloging in Publication Data
Chapanis, Alphonse.
The Chapanis chronicles: 50 years of
human factors research, education, and design
p. cm.
Includes bibliographical references
ISBN 0-9636178-9-3 (hc) : $34.00
1. Technology. 2. Human engineering.
3. Human factors 4. Engineering design. 5. Title

TA 166 C43   1999   620.8′2   99-072642

TABLE OF CONTENTS

PROLOGUE

THERE HAVE BEEN THREE major periods of retrenchment in my professional life. The first was in 1979 when I moved into my new Communications Research Laboratory before I was retired from The Johns Hopkins University in 1982. Having served on the faculty for 35 years, I had accumulated large files of correspondence, lecture notes, copies of examinations, course outlines, financial records and other documents for which I saw no future need. These were all consigned to a shredder. Several cartons of books were donated to Goucher College, but I was able to transfer some of my files and most of my books, journals, and reprints to my new laboratory.

The second was when I moved from the Communications Research Laboratory into my smaller consulting offices in 1984. Still more correspondence and records were purged and consigned to an incinerator. Several hundred books were donated to local college libraries. Enough teak and oak furniture – desks, files, bookcases – and office equipment were moved to furnish the new offices; the rest were donated to the Department of Psychology at Hopkins.

The third was when I closed my consulting offices in 1996. That was the most extensive retrenchment so far. A very few items of furniture and office equipment were transferred to my study at home. I sold some things, gave some away to close friends, and donated the rest (a truckload of furniture and equipment) to the Indian Creek Schools, a private school where Viv's youngest son is a teacher and where one of her grandsons is a student. Several cartons of books and journals were donated to the HFES Student Chapter at SUNY Buffalo. A collection of over 2000 slides amassed over the years to illustrate my lectures was donated to the University of Connecticut. Several cartons of correspondence and notes were sent to the Archives of the History of American Psychology. Hundreds of books were donated to Loyola College, an equal number of reprints were consigned to an incinerator, and many records of my consulting activities were similarly destroyed. The next period of retrenchment is still to come, and it will be the most heart-wrenching of them all. We have to dispose of most of the contents of a three-story townhouse and save only those few items that we can use in our final retirement apartment. At that time, I shall be able to keep only a few, a very few, of my most cherished books, a bookcase or two, perhaps one file, and my computer. Everything else – the reprints I still have, my complete collection of *Human Factors* and *Ergonomics* and smaller sets of other journals, correspondence – all this will have to go.

I could not think of facing that daunting task without trying to capture at least the highlights of those documents still in my possession so that in my retirement years I may read about them and reminisce over the accomplishments and disappointments of my career. So this book has been written primarily for myself. But I have also written it for my children, grandchildren, and, hopefully, great grandchildren so that they may read about what I did and what I thought about what I did. It will be an unexpected bonus if a few historians will be interested enough to read it for the small part I played in the development of human factors.

This has been a difficult book to write. I was repeatedly distressed to find how much of the past I had forgotten, and with so many of my records gone I had no way of recapturing it. Then, too, in rummaging through those files and records I still have I would find something – a reprint, a note, an entry in a diary, a stamp in a passport – that would trigger my memory of a forgotten event. That meant backtracking to insert the newly recovered incident or event in text that I once thought was complete. I lost count of the number of times I have gone back and forth in writing this book.

Rather than describing events in their strict chronological order, I have organized the book into chapters each of which is a more-or-less coherent group of related items. As a result, events in one chapter may have occurred coincidentally or even before events in some other chapter. Although unconventional this organization was congenial to my style of writing. Insofar as possible I have tried to document everything I have written here and I have admitted honestly when my memory failed me. But at age 81 I don't think any apologies are necessary.

Another quite different perspective on my professional life is contained in the 80-page transcript of the Oral History of Alphonse Chapanis, Ph.D. recorded on May 27 and June 30, 1986 for the American Psychological Association. Although the oral history supplements what I have written here, it lacks the organization and precision of this book because it contains no citations, quotations, or exact references to other records. Instead, it consists of unrehearsed responses to questions put to me by my interviewer and former student, Gerald P. Krueger. But if you want to know what I really thought and felt about things, the Oral History is the document to read.

A. Chapanis
15 August 1998

1. INTRODUCTION

My first article was published in 1937 (Chapanis, 1937). As I look back at the 60+ years that have elapsed since then, I see a professional life filled with teaching, lecturing, traveling, researching, publishing, and consulting. Sprinkled throughout are some accomplishments tempered by major disappointments and failures. But I also find disturbing evidence of serious lapses of memory. For example, I find among my papers reference to an opening address I gave at a joint International Ergonomics Association/Ergonomics Society conference on "Ergonomics and Job Satisfaction" at Fulmer Grange, Slough, England, in September 1977. Although my passport confirms that I did indeed visit Great Britain at that time, I have absolutely no other record or recollection of that meeting.

Nor do I have copies of invited addresses and sometimes no clear recollections of addresses I gave at the International Symposium on Practical Applications of Ergonomics in Industry, Agriculture and Forestry in Bucharest, Rumania, in September 1975; an invited address in the Department of Applied Psychology, University of Zürich, Switzerland, also in September 1975; the

NATO Symposium on University Curricula for Ergonomics/Human Factors Engineering in Berchtesgaden, West Germany, in March 1976; the USSR Academy of Sciences in Moscow, USSR, in June 1977; the IBM Systems Research Institute in New York City, in February 1979; and the more than a score of other meetings, symposia, and conferences in the United States and Canada, in Europe (Belgium, Britain, Czechoslovakia, France, Finland, Germany, Greece, Italy, Yugoslavia, The Netherlands, Norway, Poland, Switzerland, USSR), in the mid-East (Israel), in Asia (India, Thailand, Japan) and in the South Pacific (Australia) for which my diary and expired passports provide fleeting but mute testimony. So in the twilight of my life I have undertaken to commit to paper those memories I still have before they have all faded away entirely and those records I have not yet destroyed.

Many and diverse springs in this country and abroad contributed to the stream and then the river that we call human factors or ergonomics (Parsons, 1972; Roscoe, 1997). But mine is a personal history – a kind of autobiography – since it is concerned almost exclusively with what I have done and events that touched my life. Still, I have been involved with human factors for more than half of my entire life and I would like to believe that some of the things I have done have helped to shape the profession. Despite its very personal flavor, my account may have some historical value because much of what I write about exists only in unpublished reports and other documents that are difficult or impossible to find. Then, too, time has so decimated the ranks of those of us who lived through the infancy of human factors that in some cases I am the only person surviving who can write about them. Finally, my records may help to correct some inaccuracies that I have seen in other historical accounts. So, I make no pretense that this is a complete history of the development of human factors since its beginnings. It is, rather, a personal account of what I did and what I thought about what I did. For that I make no apologies.

I START WITH YALE

My graduate training at Yale was fairly conventional for those times and included such courses as History and Systems, Quantitative Methods, Physiological Psychology, and Learning – taught by some of the most eminent psychologists of that period: Donald G. Marquis, Clark K. Hull, Robert M. Yerkes, Leonard Doob, Carl I. Hovland, Arnold Gesell, Walter R. Miles, and Mark May. But World War II, more than anything else, determined the direction my professional life would take. In my last years at Yale, Walter Miles had begun to do research for the National Research Council on a practical problem about which our armed forces needed to know more – dark adaptation and night vision. His research led to the use of red lighting for the preservation of night vision – a discovery for which he later received the Warren Medal from the Society of Experimental Psychologists.

2. THE WORLD WAR II YEARS
1942-1946

Professor Walter Miles was instrumental in my appointment as a psychologist in the Aero Medical Laboratory (AML; now the Armstrong Laboratory), at Wright-Patterson Air Force Base, Dayton, Ohio, in the fall of 1942, the first psychologist to be employed there. I was not a civilian for very long, however, because immediately after I received my Ph.D. in absentia from Yale in February 1943, I was commissioned a second lieutenant, Air Corps, and sent to the School of Aviation Medicine, Randolph Field, Texas, for training as an aviation physiologist. That training completed, I was reassigned to AML in May 1943 and, although I traveled extensively in the following years, this remained my home base throughout the war years.

THE MISSION OF THE
AERO MEDICAL LABORATORY

Although the AML was a medical laboratory, it was located on a base that was responsible for the design, testing, and specification of the full gamut of devices and equipments (from sunglasses to aircraft) needed by the Air Force. The primary mission of AML was to ensure that every item of equipment was suitable for and adapted to the personnel who were to use it. Research at the laboratory was at the forefront of physiological and psychological science, investigating phenomena – hypoxia, aeroembolism, high-g forces, explosive decompression, night flying – that at the time were poorly understood. Confronted by the urgency of providing immediate solutions to practical problems there was little time for basic research in those war years. Still I feel that I was able to make some modest contributions to the overall mission of the laboratory and the Air Force.

PILOT ERROR?

It was at AML that I was asked to figure out why pilots and copilots frequently retracted the landing gear instead of the landing flaps after landing, a problem that Fitts and Jones (July 1, 1947)[1] documented so thoroughly years later. I recall reading dozens of accident reports that concluded these accidents were caused by

---

[1] Many citations in this book are to unpublished reports, memoranda and documents that may be difficult or impossible to find now. In referring to them I have not used any standard form of citation, but have tried to provide as complete and accurate information as possible about them. In particular, I cite dates exactly as they appear on the original documents.

"pilot error." What I found on inspecting the cockpits of planes such as the B-17, the "Flying Fortress" of those years, was two identical toggle switches side by side, one for the landing gear, the other for the landing flaps. Given the stress of landing after a combat mission, it is understandable how they could have been easily confused. I have called this "designer error" and not "pilot error." The ad hoc remedies proposed at the time (separate the controls and/or shape code them) were substantiated in the human factors literature years later. Another remedy was a more mechanical one – installing a sensor on the landing struts that detected whether they were compressed by the weight of the aircraft. If so, a circuit deactivated the landing gear control in the cockpit.

A TEST OF LUMINESCENT MATERIALS
FOR COCKPITS

My first documented contribution was a minor, almost trivial one. The Materials Laboratory, Experimental Engineering Section, had submitted to AML samples of four luminescent surfaces, an orange-red, two yellow, and one green, that were proposed for use in aircraft cockpits. The question was this: What effect did these materials have on dark adaptation when they were viewed under ultra-violet light? My tests were straightforward (Chapanis, November 4, 1942). I concluded that the differences were small but that the orange-red sample affected dark adaptation less than the green one.

A SPECIFICATION
FOR EMERGENCY FLARES

My second contribution was a little more substantial. It was a memorandum I prepared about design factors that were important for optimum visibility and detectability of emergency flares. In it I reviewed available evidence and made recommendations about such variables as the size of the flare, its intensity, color, flash duration, and eclipse period. Reading it today, I find it a competent job for the time. Although I wrote the memorandum, it was signed by the Chief of the branch to which I was assigned (Gagge, January 9, 1943). Major Gagge and I tested some of my recommendations one night on a golf course near Dayton, and flight tests were later made from a B-24 flying over sections of Lake Michigan in which test flares had been dropped.

DEVELOPMENT OF
ANOXIA DEMONSTRATION CHARTS

Another product I developed was a simple device for high-altitude indoctrination. Before World War II most commercial and military flying was done at moderate altitudes. In 1938, however, B-17s, the forerunners of the "Flying Fortresses", were the first high-altitude aircraft to be used in routine service – non-combat training, transport, and messenger and liaison duties. Their successors, B-17Cs (Fortress 1), purchased by the Army Air Corps in 1940, were the first to be used in combat. Both these aircraft had service ceilings of over 35,000 feet, altitudes at which the air was so thin that pilots could no longer survive without supplemental oxygen. Pressurized aircraft were still some years in the offing.

These aircraft and other high-altitude aircraft that were developed soon after were so new that many pilots, especially some older macho-types, resisted using oxygen because they seemed to regard it as a sign of weakness. For that reason, universal high-altitude indoctrination was instituted early in the war and continued throughout its duration. It is still done today. For those indoctrination sessions we needed some simple but effective way of convincing indoctrinees that breathing 100% oxygen at even moderate altitudes improved their abilities to function.

With the cooperation and assistance of Matthew Luckiesh of the Lighting Research Laboratory of the General Electric Company, Nela Park, Cleveland, Ohio, I developed two versions of an Anoxia[2] Demonstration Chart (Chapanis, 1946a). Both are variations of the Visibility Indicator designed by Luckiesh and Frank K. Moss. Type AAF-1 consists of text and AAF-2 of numerals, both on backgrounds that shade from white at the top to black at the bottom. In practice, trainees were taken to a simulated altitude of 15,000 or 16,000 feet in an altitude chamber without supplemental oxygen. After a few minutes they noted how far down the chart they could read. After they donned oxygen masks and started breathing 100% oxygen, the charts typically appeared much brighter almost immediately and they could read further down the charts. In 1986 when I visited an altitude indoctrination unit at Wright Field during an annual meeting of the Human Factors Society I was astonished and gratified to find my device still being used.

---

[2] Hypoxia is the more medically correct term.

2. The World War II Years – 1942-1946

STUDIES OF DARK ADAPTATION
AND NIGHT VISION

A substantial part of my work was on problems of dark adaptation and night vision.[3] [A number of reports on these problems are not worth documenting here.] We studied ways of testing night vision with various devices that had been proposed, after which we developed a small portable night vision tester of our own design (Pinson & Chapanis, 1945). Most investigators at that time took it for granted that a dark adaptation test, the faintest light one could see in the dark, was a valid measure of ability to see at night. Our device, however, was essentially an acuity test. It measured a person's ability to discriminate forms at very low light levels. Tests of this device at Fort Knox were so successful that it was adopted by the Armored Forces because of its low cost and simplicity. In June 1945 it was also adopted for use in the AAF when more elaborate test devices were not available.

Debunking Some Russian Claims

A number of reports from Russian laboratories (Kekceev[4], 1937; Kekcheev, 1943; Kektcheeff & Astrovsky, 1941; Lasareff, 1937; Lasareff & Dobrovolskaia, 1937; Streltsov, 1944; Turugin, 1937) claimed to show enormous effects of intersensory stimulation on dark adaptation and night vision. For example, Kekcheev (1943)

[3] A complete bound collection of the reports I wrote at AML has been deposited with The Archives of the History of American Psychology, The University of Akron, Akron, Ohio 44325-4302.

[4] Kekceev's name is variously transliterated. I use the spellings as they appear in the original articles.

19

stated that "Several months ago we experimented in expediting adaptation by means of light muscular exercise... Experiments made on ten subjects with the help of the adaptometer revealed that it was possible to reduce the period of adaptation from 25-45 minutes to 5-6 minutes." Other articles stated that auditory, olfactory, gustatory, labyrinthine, thermal, pain, tactile, proprioceptive and interoceptive stimuli were all effective in changing the sensitivity of the dark-adapted eye.

The Russian reports were, however, extremely skimpy in presenting details of apparatus and experimentation. So in 1943 I tried unsuccessfully to get more detailed information about these studies through military and diplomatic channels. After I left the Air Force and was no longer constrained by military regulations, I tried to communicate directly with Dr. Kekcheyev through the mails. My letter to him was returned unopened from Moscow.

In any case, the Russian reports were so insistent and the claims so dramatic that in the fall of 1943 and spring of 1944 I directed a series of three experiments on dark adaptation, low contrast sensitivity, and form discrimination at very low light levels, with and without various forms of intersensory stimulation. The conclusion was stated simply, "The results of all experiments are completely negative. None of the stimuli...either facilitated or inhibited dark adaptation, contrast sensitivity, or form discrimination at low light levels" (Chapanis, Rouse & Schachter, 1949).

A Novel Way to Get Your Vitamins

One of our missions at AML was to evaluate ideas and devices submitted to the Air Force by well-intentioned persons. One I recall vividly was a beta-carotene inhaler which its author claimed greatly enhanced one's ability to see at night, although he provided no data to support his claims. I don't think he was in any way influenced by the Russian work I have cited above. More probably he was trying to exploit the fairly recent discoveries (this was in the early 1940s, remember) of the importance of Vitamin A in the regeneration of rhodopsin. At any rate, tests I made with subjects inhaling beta-carotene showed no improvement in ability to see at night. Its inventor, motivated I feel less by altruism than by his inability to convince us to buy hundreds of thousands of inhalers for the Air Force, then took his device to his congressman who demanded still further tests. Our commanding general supported my research findings, however, and the furor gradually subsided.

A Temporary Assignment to Fort Knox

Our work on night vision in the Air Force was considerably more advanced than that in the Armored Forces, so when our ground forces began fighting at night in the Pacific campaigns, I received orders to report to Fort Knox, Kentucky, on May 16, 1945, for one month "to assist the Armored Medical Research Laboratory in establishing a night vision training program for the Army Ground Forces." There, in addition to the usual lectures, demonstrations and tests, we laid out a walking course through varied terrain and took platoons of officers out in the middle of the night to walk the course without any artificial illumination. At the beginning of these sessions, trainees stumbled about a great deal, but at the conclusion they were able to walk the course with greater confidence. When I returned to AML I wrote a 27-page report (Chapanis, 1 July 1945) describing the night vision testing and training program in the ground forces with some implications for our own programs in the Air Force.

One experience during my Fort Knox assignment was both fun and instructive. I told the officer to whom I reported that I would like to drive a tank. He arranged it one morning and sat at my side to instruct me in how to maneuver it. When I further confided that I would like to knock down a tree, he pointed to one a little smaller in diameter than a telephone pole, told me how to swivel the turret and cannon out of the way, and told me to have a go at it. My first assault only succeeded in shaking the tree. My mentor's advice was to back up, gun the motor and really slam into it. My second attempt succeeded in breaking the trunk and my third resulted in flattening it. What a feeling of power! But I also came away with a profound respect for the men who had to fight in these vehicles. Visibility was severely limited, the ride over rough terrain was bone-rattling, and the noise deafening.

In Summary

Our work on night vision and night flying was so successful that I was invited to write two general articles about vision and visual problems for distribution throughout the entire Air Force (Chapanis, 1945, 15 March 1945). It is interesting to observe that, however successful this work was at the time, it has little relevance to most military activities today. Modern technology has produced infrared and low light level amplification systems that enable one to see even in total darkness. So, as has happened in other instances, technology rendered obsolete much of the good work that had been done on dark adaptation and night vision. It has not all been wasted, however, because I believe that some of these principles are still applicable in the submarine service and, in the civilian arena, for night driving.

VISUAL PROBLEMS
IN THE DESIGN OF AIRCRAFT

Late in 1942 pilots frequently became sick when they flew the A-30, a low-level attack bomber. Although several hypotheses were advanced to account for these problems, noxious gases in the cockpit being a major one, our final conclusion was that the pilots were becoming nauseated because of optical distortion in the windscreen. During low level flight the horizon and other objects on the ground danced about so haphazardly that pilots were in effect becoming airsick in a manner akin to seasick. I later reported on other distortion problems in the B-25, a two-engine medium bomber (Chapanis, 25 November 1943), and the B-29, a four-engine, high-altitude bomber (Chapanis, 7 May 1944). I also participated in mock-up inspections of these aircraft and of the XF-12. My recommendations were ultimately incorporated into the Handbook of Instructions for Aircraft Designers.

I was never satisfied with my work on distortion because I was left with a feeling of incompleteness, of a problem not really solved. We needed an objective way of measuring and specifying distortion quantitatively, and I was never able to find any satisfactory way of doing that. In addition, I was never able to find a way of measuring the effect of distortion on visual functioning. Distortion made bombardiers miss their targets, it made crews airsick, and it was a source of many complaints, but I could never find a way of measuring and quantifying those effects in a way that was convincing to engineers and designers. The Howard-Dolman test of depth perception which I used extensively showed effects, but the test lacked face validity, and so this whole area of research was one of my major disappointments.[5]

Measuring Fields of View

Fields of view (through goggles, equipment worn on the face, canopies) was another area of research and here I was more successful. For that purpose I prepared a mathematical procedure for integrating visual field limits into a single measure of the size of the cone of vision in steradians, or multiples of a unit solid angle (Chapanis, 12 May 1944). I recall seeing this measure used some years later in some non-military context, but I can't recall where or when.

I put this technique to use in measuring and comparing the pilots' fields of view from 11 US, British, Japanese, and German fighter aircraft. In another study the downward (over the nose) visibility for five US fighter aircraft was almost perfectly correlated with the rate of taxiing accidents per 100,000 hours of flying time. Those aircraft with the most restricted downward visibility had the highest rates of taxiing accidents (Pinson & Chapanis, 1946).

SOME MISCELLANY

Another device submitted to us for test had a rotating sector that alternately exposed the right and left eyes. When used with

---

[5] Having had 55 years to ruminate over these problems I think I know now how I could have successfully solved both of them.

the eyes parallel to the line of flight, its inventor claimed that it should greatly improve one's depth perception. The idea was that the movement of the aircraft effectively increased the interoccular distance and so increased one's ability to see variations in depth on the ground. We tested this device while viewing the ground through the open bomb-bay doors of a B-25. Although there was a jnd (just noticeable difference) of improvement in depth perception, we concluded that the device was impractical for tactical use.

We also had the responsibility of evaluating sunglasses and goggles for the Air Force. We did not actually design eye protective devices, although we did submit recommendations about design features to their manufacturers. One successful sunglass design is described in an article by Pinson, Romejko and Chapanis (1945).

THE WAR ENDS

VJ Day on August 15, 1945, marked the end of hostilities and my time after that was devoted to completing projects that had already been started, examining and testing captured German and Japanese equipment, and translating captured German documents. On 30 November 1945 I had been promoted to Captain and shortly thereafter (I can't find the exact date in my records) I was appointed Chief of the Vision Unit, Biophysics Branch, of the laboratory. My staff at that time included First Lieutenant Richard O. Rouse, Jr. (psychologist), Technical Sergeant Max Cohen (aerial gunner), Sergeant Edwin M. Siegel (physicist), Sergeant Stanley Schachter (psychologist), Sergeant Mortimer Marks (plastics manufacturer), Corporal Marvin Schondorf (medical student), Mrs. Rose Stump (engineering aide), Mrs. Mary Hall (engineering aide), and Miss Patricia J. Duffy (secretary). Two other officers, Captains Richard C. Armstrong and Walter J. Romejko (both ophthalmologists), had also served in the Vision Unit, but I believe they both had been successful in obtaining early discharges and so had never served under me.

I was never enamored of military life. One source of my discontent appears in the Preface to the volume of my collected reports. That Preface written 1 September 1946 complains about "...a large number of monthly Project Reports, Semi-Annual Project Reports, and Annual Project Reports; Inter-Office Memoranda; Conference Notes; and Army Air Force Specifications which I prepared as part of my duties. It seems to me now that far too much of my time at the Aero Medical Laboratory was spent in writing and far too little time was spent in the laboratory or field procuring the data necessary to make all this writing meaningful and well-founded in fact."

A Psychology Branch was established at AML on 1 July 1945 and Walter F. Grether II "...was one of the first of the staff of the new branch to arrive at AMRL in August 1945. Dr. [Paul M.] Fitts arrived a few days later" (Green, Self, & Ellifritt, 1995, p. 1-9). This branch, now called the Paul M. Fitts Human Engineering Division, has over the years become one of the most, if not the most, successful and foremost human engineering laboratories in the US. I'm not sure now why I never joined Fitts' unit. Whatever my reasons, when I was offered a position with The Johns Hopkins University late in 1945, I jumped at it and requested a release from military service at the earliest opportunity. According to my military records I left Wright Field on 3 June 1946.

SUMMING IT ALL UP

My work at the Aero Medical Laboratory has had no major or lasting impact on science or technology and only some parts of it could be classified as human factors research and development as I define it. My work on vision and visual problems did, however, bear fruit years later. In 1952 I was selected to chair a conference on "Chart reading under red illumination" at the Naval Medical Research Laboratory, New London, Connecticut [about which more later], and in 1953 I was one of the four authors of a 42-page report on visual reconnaissance [also described later].

In the September 1, 1946, Preface to my collected AML reports, I wrote, "In retrospect, a few of the studies appear good, most are mediocre and some are poor. The war years were, for many reasons, a difficult period in which to do research. Adequate library facilities were not available to assist in the preparation of scholarly reports. Time was short and military authorities placed great emphasis on the accumulation and dissemination of data to the services in the shortest time practicable. Military authorities in position of higher command, moreover, were frequently not interested in the data per se but rather in the implications and applications of these data. For this reason, it has been necessary in some instances to make specific recommendations on the basis of scanty evidence. I recognize full well the dangers of such extrapolations and can say only that the choice of phraseology was frequently not mine. Finally, many of the reports were "command" projects in the sense that they were assigned by higher authority."

From my perspective today with perhaps somewhat more maturity and wisdom, I think that at the time I was still too thoroughly imbued with academic values and that the underlying

27

philosophy of those times was correct. There *was* a war to be won. The niceties of elegant research were correctly subordinated to the question, "How can we use this information to make or do something better?" Despite my criticisms and complaints at the time, I feel now that the war years were not wasted. Questions of diverse sorts were asked and I did the best I could to answer them with the time, resources and skills available to me. I would like to feel that the contributions I made at that time helped in some small way to further our war effort. Perhaps the most important outcome of this war-time experience, however, was that it made me acutely aware of how many immensely important problems there were to be solved in the world, how much more difficult it was to work on these problems than on the abstract problems of the academic laboratory, and how inadequately my academic training had prepared me to deal with problems of such complexity. Those lessons have stayed with me and have been integrated into my research philosophy.

3. SYSTEMS RESEARCH
1946-1958

The war over, I left the Air Force to join The Johns Hopkins University's Systems Research Field Laboratory on the Beavertail Point (Conanicut Island) Naval Training Facility in Narraganset Bay, Rhode Island. I'm not sure exactly when I joined the laboratory, but Progress Report Number 2[6] , dated 1 July 1946, states that two psychologists had been added to the staff since May 1. I am surely one of the two (I believe Neil R. Bartlett was the other) so I must have joined late in May or in June. A further bit of corroboration is a typewritten copy of a talk I gave on "The function of the psychologist in the Systems Research project" to members of the Naval War College, Newport, Rhode Island, at a meeting on the 24th and 25th of July 1946. (Morgan lost no time putting me to work!) Since that talk was a fairly thorough description of

---

[6] I have donated to the Johns Hopkins University, Baltimore, Maryland, ten bound volumes comprising a complete collection of progress Reports and Research Reports, except for three highly classified research reports and one from the Motion and Time Laboratory of Purdue University. They are cataloged and stored among the Special Collections of the Milton S. Eisenhower Library.

problems psychologists could tackle on this project, I must have been associated with the laboratory long enough before that to have learned something of its activities. The talk was never published and, so far as I know, never reproduced in any other form; I believe I have the only copy.

In February 1947 I moved from the Field Laboratory to the University in Baltimore where, except for two leaves of absence, I remained until my retirement on June 30, 1982.

THE ORIGIN OF SYSTEMS RESEARCH

The following account of the origin of Systems Research is an edited version of material in the first progress report, dated 1 May 1946. In January 1945, The Bureau of Ships (BuShips), with the endorsement of other interested bureaus and naval activities, requested the National Defense Research Committee (NDRC) to conduct motion-and-time studies of Combat Information Centers (CICs) and to establish a field laboratory in which CICs could be set up and studied in a manner simulating operational conditions. These requests were rated urgent; the work was to proceed with all dispatch toward recommendations that could be put into practice in the shortest possible time.

NDRC's contractor, Harvard University, with the cooperation of the Radiation Laboratory and the assistance of BuShips, quickly assembled personnel for the work from existing NDRC laboratories and outside institutions: Psycho-Acoustic Laboratory, Electro-Acoustic Laboratory and Radio Research Laboratory, all at Harvard; Underwater Sound Laboratory, Columbia University; Motion-and-Time Study Laboratory, Purdue University; and Industrial Engineering Department, New York University. The group assembled consisted of specialists and research workers in applied psychology, motion-and-time engineering, mathematics, physics, sonar, radar, communications and countermeasures.

The Field Laboratory

During the spring of 1945, a group began studies of CIC operations on ships undergoing shakedown or refresher training under Commander Operational Training Command, Pacific Fleet, San Diego. At the same time, construction of the Field Laboratory got underway at the Naval Training Facility, Beavertail Point on Conanicut Island in Narragansett Bay, Rhode Island. By June, some operational results had been obtained and the Field Laboratory was completed. In July 1945, major items of equipment at the Laboratory were in operation and some experimentation could begin.

The Systems Research Field Laboratory was unique in several respects. It provided facilities for mocking up and evaluating realistically the operation of any CIC, as well as the many ship and fire control stations associated with CICs. A central area of the laboratory (one thousand square feet in size) had movable partitions suspended overhead and removable floor panels that made it possible to reproduce quickly any size or shape of CIC. The mobile floor panels covered cable wells that contained electrical, sound power, interphone, and radio outlets so that any piece of equipment could be quickly relocated and put into operation. Another large room adjoining the central laboratory contained target generation, control, and recording equipment.

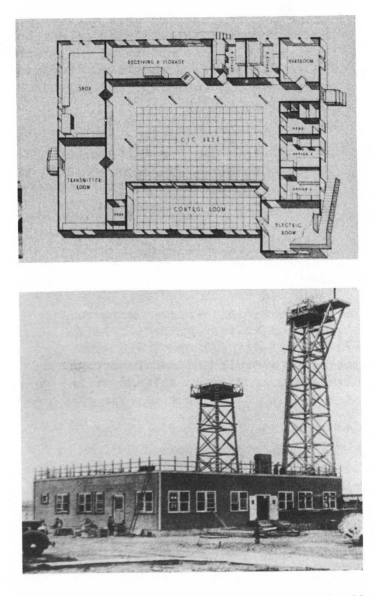

Figure 1. The Systems Research Field Laboratory, May 28, 1945. The bottom photograph shows the exterior of the laboratory and radar towers. The top illustration is a plan view of the interior.

Changes at Wars End

Hardly had the laboratory gotten underway when the war came to its sudden end. Personnel scattered quickly to new positions or returned to graduate schools and prewar positions. Many experiments that had been planned had to be abandoned, some were brought to a stop, and only a few were carried through to completion. By late fall of 1945, the laboratory's research staff had dropped to less than 10 percent of its former strength, and the engineering staff was almost as severely depleted. Under great pressure and handicaps, the remainder of the Harvard staff managed to get reports of its work together and to bring the program to a close. Fifteen reports cover this phase of the work. One of those reports (Systems Research Laboratory, November 1, 1945) describes work done on the redesign of the CIC in the USS Louisville (CA-28), a heavy cruiser. It is one of my favorite examples of a link analysis and I have used it in two of my books (Chapanis,1959, 1996b) and in numerous lectures.

THE JOHNS HOPKINS UNIVERSITY ACQUIRES
SYSTEMS RESEARCH

At the request of BuShips and other interested branches of the Navy Department, the Office of Research and Inventions (ORI) undertook to provide for a continuation of systems research on a peace-time basis. The contract with The Johns Hopkins University was placed in effect on December 1, 1945, and, on January 1, 1946, responsibility for the Field Laboratory (as well as those personnel able and willing to continue with the work) was transferred to

Johns Hopkins. The transfer of all activities from Harvard to Hopkins was finally completed on April 30, 1946.

Navy Sponsors of Systems Research

Systems Research operated initially under Contract N5-ori-166, Task Order 1, between the Office of Research and Inventions, U.S. Navy, and The Johns Hopkins University. According to Progress Report Number 2, dated 1 July 1946, Navy sponsorship was changed to the Special Devices Division, Office of Research and Inventions, U.S. Navy. A further change to Special Devices Center, Office of Naval Research, appears in Progress Report Number 4, dated 1 November 1946. Progress Report Number 26, dated 1 February 1952, shows the project under the Systems Coordination Division, Naval Research Laboratory, Office of Naval Research. Finally, Progress Report Number 30, dated 1 February 1953, listed the project simply under the Office of Naval Research, where it remained until the termination of the contract on October 31, 1958. Because the progress reports mentioned or discussed military equipment, they were all labeled RESTRICTED, a security classification that was eventually canceled.

Directors

The systems research project, the first of its kind at a university, was administered by Clifford T. Morgan, newly appointed professor of psychology, but its associate director, Ferdinand Hamburger, Jr., was a professor of electrical engineering. Directorship of the project changed several times. When I joined the Systems Research Field Laboratory in the summer of 1946,

Morgan was director, a position that he gave up on July 1, 1949, to devote full time to his activities as chairman of the Department of Psychology. Wendell R. Garner assumed the position vacated by Morgan, but when Garner began a year's leave of absence on September 1, 1952, Morgan and I became joint project directors. I resigned on April 6, 1953, to leave for a year at the Bell Laboratories in Murray Hill, New Jersey, leaving Morgan as sole director. When Garner returned from a year's leave of absence, he assumed directorship of the contract on July 1, 1953, and Morgan resigned at the same time. I took over as director on May 1, 1955, and continued in that position until the termination of the contract on October 31, 1958. Ferdinand Hamburger, an associate director of the project since its inception, resigned on October 1, 1956.

Goals of the Project

No long-range or basic research had been planned or attempted under the war-time contract with Harvard. Its peace-time successor at Hopkins planned a broader and more basic approach to information systems and provided for "...physical, psychological and time and motion studies of military information systems and of the various devices for the indication, display, control and transmission of information, which make them up." The following were the objectives of the project, and I quote from the first progress report, dated 1 May 1946:

1. To conduct basic research in problems of human perception and performance, out of which will come suggestions and recommendations for the improvement of devices so that they will best suit the capacities of their operators;

2. To investigate the psychological and time-and-motion factors which are important in the most effective use of informational systems;

3. To conduct tests in which the devices are appraised (in the light of man's ability to use them under normal operating conditions) for the kind, amount, and accuracy of information which they can handle;

4. To set up and test layouts of present or proposed information systems, to measure their performance, and to propose possible methods of improving them.

Subcontracts

The project also had subcontracts with the Industrial Engineering Laboratory of New York University, the Motion-and-Time Study Laboratory and Applied Psychology Laboratory of Purdue University, and the Psychophysical Research Unit of Mount Holyoke College.

In my opinion, except for the link analyses, the motion-and-time engineers who worked on the project always seemed to be floundering and never seemed to be able to come up with studies that really attacked important problems. The reason, I think, was that our problems were not primarily simple ones of repetitive movement and motion economy. They dealt with *information*, a more nebulous commodity which the procedures and techniques of motion-and-time engineering could not handle. Moreover, motion-and-time engineering relied primarily on simple observational techniques that did not equip them with the expertise needed to design meaningful experiments of the kind the project required. So the subcontract with New York University was terminated in the summer of 1947.

Of the two laboratories subcontracted to Purdue University, the one with the Applied Psychology Laboratory was the more useful due primarily to the work of Robert B. Sleight. For example, I have seen one of his papers (Sleight, 1948) cited in the current human factors literature. When we terminated the subcontract with Purdue in 1948, Sleight left Purdue and came to work with us, where he continued to be productive. The Mount Holyoke group did produce a number of studies, but they were so basic they had little relevance to any applied problems. That subcontract was terminated on June 30, 1950.

Supporting Personnel

In addition to the academics, the field laboratory had a complement of Navy enlisted men under Commander J. S. Wylie, a career naval officer. They rounded out a rare combination of talents. The Hopkins engineers provided the technical expertise necessary to operate the complicated systems under test. They also designed and constructed devices needed for experimentation. Commander Wylie brought to the work his operational skills and expertise, and for subjects we had the Navy enlisted men, all of whom had been trained in the use and operation of the devices we tested.

Systems Research Moves to Baltimore

Despite its unique features, the field laboratory suffered from its isolated location. Research was expensive because it required a completely separate plant and facilities far removed from Baltimore. Moreover, keeping the isolated laboratory staffed with competent personnel proved a difficult task. Accordingly, in June 1948, we moved the field laboratory and all its equipment to a large building (long since demolished) located at 1315 Saint Paul Street in Baltimore. That building had been taken over for the university's contract research activities under a new administrative entity called the Institute for Cooperative Research (ICR), an organization that was later replaced by the University Sponsored Projects Office. With that move we lost all our Navy personnel.

The psychological laboratory, as it would now be called, had over 7,000 square feet of floor space and twenty-nine rooms for offices and laboratories. The new laboratory also had at its center an "island" that duplicated the versatile CIC area that had been so successful at the field laboratory. As had been the case in the field laboratory, movable floor panels concealed electrical and other connections, and the interior of the island could be quickly modified as required. Research continued at 1315 Saint Paul Street until December 1954 when all activities were moved to Ames Hall on the Homewood Campus of the university.

SOME PRODUCTS OF SYSTEMS RESEARCH

In his historical review, Parsons says about the systems research project, "The Johns Hopkins program was indeed prolific" (Parsons, 1972, p. 113). And so it was. Studies were conducted not only at the field laboratory, but in the electrical engineering and psychological laboratories at the university in Baltimore. In the roughly thirteen years of its existence, 221 reports had been prepared under the auspices of the project, of which 143 (nearly two-thirds) were published in the open periodical literature. Many of these reports are classics, providing the basic data for human factors design guides that are the current bibles of the profession. Some reports, roughly 25 percent, were evaluations of specific radars and other equipment requested by one or another branch of the Navy. Although these reports were useful to the agencies that requested them, they had no survival value because the equipment of those days has long since been replaced by much more sophisticated devices.

Contributions of the Engineers

Although my interests are concerned with psychological and human factors studies, my work was strongly dependent on contributions from the engineers, and they were substantial. The engineers not only maintained all the radars we used, but they designed and developed movable target generation equipment to simulate targets realistically on the radars. Equally important were the many specialized equipments they designed and constructed for our experiments. I briefly describe two of several that directly involved my own work.

A bearing counter. In the middle 40s target bearing and range information on virtually all radars was obtained by rotating a cursor on the plan position indicator (PPI) of the radar to intersect a target. Bearing information was read from the intersection of the cursor with a scale that surrounded the PPI. Range information on most radars was obtained by positioning an electronic "pip" over the target and reading the range from a counter. A simple thought occurred to me – why not have bearings read from a counter? There were, however, no suitable counters available for this purpose because some difficult requirements had to be met by such a device:

(1) it had to operate freely in both directions, (2) it had to operate smoothly, (3) numbers had to click into place, and (4) numbers had to increase by units from 000 to 359 and then to 000 again, or, in the reverse direction, decrease from 000 to 359 and then back to 000 again.

Mr. William A. Hamilton in our engineering department designed and constructed such a bearing counter. Figure 2 shows the experimental bearing counter mounted on a VJ radar. Its design is detailed in an appendix to one of my reports (Chapanis, 1 August 1947).

Figure 2. Our experimental direct reading bearing counter (A) mounted on a VJ Remote Radar Indicator; (B) is an experimental placement of the range counter; (C) is the bearing scale; and (D) is the conventionally located range counter. (From Chapanis, 1 August 1947.)

The bearing counter was tested in three separate studies (Chapanis, 1 November 1946; 1 August 1947; 20 June 1949) and in all three the results showed that with it target information could be obtained faster and with fewer errors than with the bearing scale. The improvement was not only statistically significant but large enough to be practically important. Although this was a simple idea, its test was made possible by a unique device that had never been previously available. Today, incidentally, all radars I have seen are equipped with bearing counters.

An autocorrelator. In the late 40s or early 50s I became intrigued with the question of how well people could generate long strings of random numbers. I had 12 subjects, six statistically naive and six statistically sophisticated, write out in one sitting 2,520 numbers in what they thought was a random order. Collecting the data was easy; the problem was how to analyze them. Among other things, I wanted to get some measure of serial dependencies – how much what a subject wrote was influenced by what he had already written (autocorrelations, in short). To try to compute autocorrelations on such large sets of data with the equipment available was so forbidding that I at first gave up all attempts to do so.

Our engineers came to my rescue. In 1951 and 1952 they designed and built what I believe is the first autocorrelator. It was an imposing device, a large rack on which were mounted over a hundred relays and impressive associated circuitry. Raw data were punched into tape and read by a tape reader. Tallies appeared in digital readouts. Each lag required a separate run through the computer. Now, of course, the same computations can be performed with devices as small as a hand-held computer. To put our calculator into perspective, however, the first true electronic computer, ENIAC, had been completed just five years earlier. Its successor, which replaced electromagnetic relays with 18,000 vacuum tubes designed for radar and television, came a year later and filled a very large room. So, for its time, our autocorrelator was an impressive invention.

With this new electronic wizard I was able to compute the autocorrelations I wanted. Once that had been done, other more pressing research intruded so that the data and analyses were buried in my files. They were resurrected some 40 years later, some additional analyses were made as an activity for my retirement years, and the results were finally published (Chapanis, 1995). When I had no further use of the computer in 1953, we either sold it or gave it away to another research organization that had requested it.

A serendipitous sequel. In 1957 I was engaged as a consultant by the Humble Oil and Refining Company, Houston, Texas, to advise about the display of seismic data. One way of exploring for oil was to place a dynamite charge in a hole in the ground and to record on sensors (geophones), spread out at various distances, the shock waves returning from deep in the earth after the dynamite was exploded. One question put to me was how best to use the seismic tracings to determine the tilt or slant of rock strata deep in the earth. Recalling my work on autocorrelating number data, I suggested that cross correlations be made of selected segments of seismic tracings and that lags among the tracings be varied continuously to find the lag that yielded the highest cross correlation. That lag should correspond to the tilt of the strata that produced the tracings.

This idea resulted in a patent (U.S. Serial Number 726,108, April 3, 1958) with Frank J. Feagin, William M. Rust, Jr., and me as authors. Figures in the patent illustrate this method of geophysical exploration, the seismic tracings that result, and the mechanisms for cross correlating tracings. Feagin was an engineer who designed the mechanisms for executing the correlations that had been programmed by Rust, the mathematician.

Since I was technically in the employ of the company when we devised this procedure, I had to assign all rights to the company and never shared in any royalties from its use by Humble or from its license to other companies. Nonetheless, I have always found this an interesting example of cross fertilization – of how work in one area of science or technology can sometimes be transferred to another completely different one.

A Digression into a Dead End

My Ph.D. thesis was on color deficiency (Chapanis, 1944) and I never really lost my love for that area of research. Even while I was conducting applied studies for Systems Research, I was at the same time doing research on color vision. From 1946 to 1953 I, either singly or with one of my graduate students, Rita M. Halsey, published ten articles on one or another aspect of color deficiency (Chapanis, 1946b, 1947, 1948a, 1949a, 1949b, 1949c, 1950, 1951a; Chapanis & Halsey, 1953; Halsey & Chapanis, 1952). Some years later I was invited to prepare an article on color vision and color deficiency for an encyclopedia (Chapanis, 1968b). I have even seen a few of these papers cited in current literature. Though this work is a digression from systems research and human factors, I describe some of it because during one period of my life it consumed a considerable amount of my time and effort.

I help scotch a fallacy. In the Air Force I occasionally came

43

across men who, although they possessed color deficient vision, were able to pass successfully the color vision tests administered at recruiting centers. My technical sergeant, an aerial gunner, was one such individual. An unmistakable protanope, he had nonetheless been accepted into the service. Some of these individuals had been "treated" by a Baltimore optometrist who claimed to have devised a way of training, or "reconditioning" color deficient persons (Dvorine, 1944a, 1944b, 1946). In essence his method required subjects to practice reading pseudo-iosochromatic charts while viewing them through colored filters. He cited numerous case studies of men who, once rejected by military services, were able to pass color vision tests after a period of training.

Although Dvorine's method was strongly criticized by some vision experts, the facts cannot be disputed. In an article (Chapanis, 1949b) I pointed out that there were at least two possible interpretations of such findings: (1) training produced a genuine improvement in color perception or (2) training taught the men how to decipher the plates by means of cues present in the plates because of imperfections or irregularities in their construction. One other fact is germane: viewing pseudo-isochromatic plates through appropriately selected colored filters enhances the contrast between figures and their backgrounds so that the figures are immediately recognizable by even color deficient persons.

Sometime in the late 40s I tested three men who had once been able to pass successfully color vision examinations at recruiting centers for military service. Two of the men had been trained by optometrists; one obtained a copy of the test and trained himself with the help of a friend. When I tested the men with a battery of tests they had never seen before, all three failed without any question. Based on this and other data I concluded: "There is, in short, no evidence here that training had any genuine effect on their color discrimination. The training was useful, however, in enabling these men to pass the tests on which they were trained by making use of cues which the color normal person never needs to use" (p. 258).

My original manuscript submitted to the editor of the American Journal of Optometry and Archives of American Academy of Optometry was rejected because the editor felt that it exposed him, the journal, and me to the possibility of a lawsuit by impugning the competence of a professional person. As I recall, I had identified Dvorine as the optometrist who had trained at least one of the men I tested. A revised, and somewhat watered-down, manuscript was later accepted and published.

I was never sure whether Dvorine was a charlatan or whether he really believed that he was able to "recondition" color deficiency. Since at that time he had a profitable practice "training" color defective persons one could reasonably suspect his motives. To be charitable, however, I am willing to accept that he was uninformed about the true nature of color deficiency and was incredibly naive. In any case, claims about training color deficiency disappeared shortly after the publication of my article, and, I'm happy to say, I have not seen any such claims made in the last 50 years.

My attempt to construct a quantitative test of color deficiency. One finding of my thesis research was that there was a progressive decrease in saturation discrimination in the blue-green region of the spectrum with increasing severity of color deficiency. That gave me an idea for constructing a quantitative pseudo-isochromatic test of color deficiency.

The first six charts were roughly spherical in form and were each composed of 439 dots three millimeters in diameter. The dots on every chart were all of the same size and of four Munsell values (3/, 4/, 5/, and 6/) in the same simple repetitive pattern. The charts had gray backgrounds (N3/, N4/, N5/, and N6/) on which were superimposed digits made up of colored dots. The colored figures were constant in hue (Munsell 2.5 BG) for all charts. They were also constant in saturation for any one chart, but three different saturations (Munsell chromas /2, /4, and /6) were used on each of two charts. So, for example, on two of the charts the figures were composed of 2.5 BG 3/6, 2.5 BG 4/6, 2.5 BG 5/6 and 2.5 BG 6/6 dots. On two other charts the figures were made up of 2.5 BG 3/4, 2.5 BG 4/4, 2.5 BG 5/4, and 2.5 BG 6/4 dots. And so on. I had selected Munsell 2.5 BG because it corresponded most closely to the neutral point of protans and deutans. Thus, an examinee was required to discriminate between (1) gray and (2) hues that differed from gray in various degrees of saturation.

The charts were constructed by my assistant, Mrs. Mary Hall, who meticulously punched the dots out of sheets of Munsell papers and then glued each dot into its appropriate position. It was a tedious job requiring precision and attention to detail because each dot had to fit into its designated position in the basic pattern. Frankly, I don't know how Mrs. Hall was able to tolerate it.

Some results. At the 1948 meeting of the American Psychological Association I reported results I had obtained with these six charts and with a criterion battery of 79 plates selected from four standard pseudo-isochromatic tests (Chapanis, 1948b). Subjects were 574 visitors to the Sesquicentennial Exhibition held in Baltimore in December 1947. The results were gratifying. The split-half reliability for my test, corrected by the Spearman-Brown

46

formula, was +0.96. The correlation between scores on my six plates and the criterion battery was +0.94. I concluded, however, that my test had fewer steps than was desirable.

An enhanced test. Eighteen new charts were then constructed following the same rationale, but with backgrounds of maximum saturation BG and digits or symbols composed of colors lying on the neutral lines for protans (Munsell 7.5 R) and deutans (Munsell 7.5 RP) but on the other side of white or gray. The colors for the protans were also one value step higher because of their well-known relative insensitivity to red. These colors had been selected on the basis of empirical tests (Chapanis, November 16, 1951; Halsey & Chapanis, 1952). Two additional charts were intended to be the start of a series for detecting rare defects of the tritan variety.

A disheartening conclusion. In 1949 or 1950 I tested 1082 subjects with an experimental Navy color vision lantern, a set of 39 plates from standard pseudo-isochromatic tests and my own enlarged set of 24 charts (Chapanis, 1950). I never reported the results on my test separately because I had meanwhile learned of work in progress under the sponsorship of the Inter-Society Color Council to construct what became known as the Hardy-Rand-Rittler charts (Hardy, Rand & Rittler, 1957) and I was chagrined to learn that they were using the same principle that I had used in constructing my charts, namely, using variations in saturation to grade degree of color deficiency. I have always wondered whether they had read my Ph.D. thesis and my 1948 report.

Although I felt, and still feel, that my test is better than the H-R-R plates, I reluctantly abandoned this whole line of work. To go any further with it, I would have had to do some extensive testing to discover whether my test was in fact better than the H-R-R test. If it proved to be, I would then have had to convince some publisher to print it. This is a formidable task as I had learned from visits to the Bausch and Lomb Optical Company, American Optical Company, and A. Hoen and Company (the company that did the lithography for the original A0 plates). Finding and preparing printing inks to match exactly the required colors, preparing gravure plates, and then printing and binding them is a tedious and expensive undertaking.

With the H-R-R plates in preparation, I thought it would have been virtually impossible to find any printer willing to undertake printing mine, even if I could show that mine was superior. I would also have had to have some extensive financial backing to do all this because I would also have had to partially underwrite the costs of printing. I did not have the financial resources to undertake such a program nor did I anticipate being able to obtain funds to do so. The H-R-R plates were constructed and tested with the resources of some large organizations and printed with the assistance of experts in color, gravure plates and printing, and ink composition. I did not have the facilities, resources, knowledge, or skills to undertake a program of such complexity and magnitude. So, with great reluctance I consigned this test to my "dead file." So many years of work that came to naught!

The original charts have been deposited with the Archives of the History of American Psychology.

Lectures on Men and Machines

Late in 1946, Morgan was invited by the Naval Postgraduate School in Annapolis to give a series of lectures on human engineering to the postgraduate engineering students at the school. He, Garner, and I wrote ten lectures, and we engaged Fillmore H.

Sanford, then an assistant professor of psychology at the University of Maryland, to convert our technical jargon into easily readable and comprehensible prose. The ten lectures, given once a week from March 25 through May 27, 1947, were such a success that we were encouraged to put them into a report. A 246-page version of the lectures (Chapanis, Garner, Morgan & Sanford, 1947) was printed just six weeks after the last lecture had been given in Annapolis. Although it had been classified RESTRICTED (a classification that was canceled in 1963) and so could be distributed only to persons who had the appropriate government clearance, the report was an almost instant success. The first printing was for 500 copies, followed by another 1,000 copies, and our supply of both printings was quickly exhausted.

Table 1. Table of contents of *Lectures on Men and Machines*

---

1. **Definitions, history and purpose**
   Objectives of lectures
   The field of psychophysical systems research
   History of psychophysical systems research
   Wartime developments
   Types of research
2. **Methods in psychophysical systems research**
   Science and technology
   The design of experiments
   Analysis of systems errors
   Summary
3. **The working environment**
   Temperature and humidity
   Oxygen deficiency
   Carbon monoxide
   Air pressure
   Acceleration

Table 1. Contents of *Lectures on Men and Machines (continued)*

   Motion sickness
   Vibration
   Noise
   Light and color
4. **Work and the work-place**
   Principles of work-efficiency
   Arrangement of controls and work-place
5. **Operation, appraisal, selection and arrangement of
      equipment**
   Operation of equipment
   Appraisal of equipment
   The selection of equipment
6. **Speech, communication and hearing**
   Factors influencing speech intelligibility
7. **Special auditory informational systems**
   Underwater sound
   Radio range signals
   Special auditory signals (Flybar)
8. **How we see**
   Visual acuity
   Eye movements
   Day and night vision
   The laws of vision
   Optical illusions
9. **Display problems in psychophysical research**
   Legibility
   Dial design
   The patterning of displays
   Design of tables and graphs
   Design of grids and range rings
   Summary of principles
10. **The future of human engineering**
   Problems and projects in psychophysical research
   Dangers and handicaps
   The Navy and psychophysical research

---

3. Systems Research – 1946-1958

We Publish the First Textbook

The success of *Lectures on Men and Machines* inspired Garner, Morgan and me to write a more general textbook, a project on which we embarked almost immediately. We struggled with the name because we did not want to alienate psychologists in our effort to convey the full flavor of what we were writing about. The result was a compromise. We called the book *Applied Experimental Psychology*, but added the subtitle *Human Factors in Engineering Design* (Chapanis, Garner & Morgan, 1949). The book was the first textbook in the field and it set the pattern for others that were to follow. It is gratifying to recall that when Division 21 of the American Psychological Association was formed it took the title of our book as its name and that when a small group of professionals got together in 1957 to form a new professional society, the name they chose, Human Factors Society, matched the first two words in the subtitle of our book.

A small but satisfying illustration. Everyone, I'm sure, has had an idea or done something that, although of no very great consequence, has left him or her smiling inwardly. One such instance appears in the first chapter of our book (Chapanis et al., 1949). I wanted to convey in some simple way the concepts of mean (constant) and variable errors in systems. Borrowing on my experiences as a marksman, I used the example of a rifleman shooting at a target and made up illustrations of various kinds of shot patterns. In one case, the cluster of shots is centered on the bulls-eye, but is widely dispersed (small mean error, large standard deviation). In another case, the cluster of shots is not centered on the bulls-eye, but the shots are all close to one another (large mean error, small standard deviation) (Figure 3). It was a simple but very effective way of illustrating these concepts. The point of the illustration was that mean errors are easily corrected and that the variable errors reflect true instability and are difficult to correct.

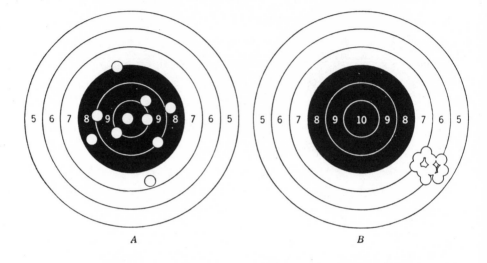

Figure 3. My way of illustrating variable and constant errors. The target pattern shot by rifleman A exhibits large variable errors and a small constant error; the one by rifleman B has a large constant error, but small variable errors. (From Chapanis et al, 1949.)

I used this illustration in an article (Chapanis, 1951b) that elaborated on these ideas and showed mathematically how mean and variable errors accumulate in systems, an article that was later reprinted in Sinaiko (1961). That illustration also appears in both the first and revised editions of the *Human Engineering Guide to Equipment Design* (Morgan et al., 1963; Van Cott & Kinkade, 1972) and has been reprinted many times. As I was writing this section in March 1998 I chuckled when I received still another request for permission to reprint that illustration (Schmidt & Lee, 1998).

Another Set of Lectures

In May, 1950, the laboratory was requested to give a series of ten one-hour lectures on Human Engineering in the Main Auditorium of the Department of Interior Building on C Street between 18th and 19th Streets in Washington, D. C. For these lectures we had written entirely new lectures and had them illustrated beautifully with slides in color created for us by professionals. The lectures were attended by hundreds of persons from government, industry, and the military services and were influential in spreading the word about research in human engineering and in demonstrating its importance to our modern technological society.

The following were the dates, lecturers, and titles of these lectures:

1. September 20, 1950, Wendell H. Garner, "Introduction and survey of the role of human engineering in Naval systems."
2. September 27, Chapanis, "The importance of human and equipment errors."
3. October 4, Jack W. Gebhard, "Human engineering factors in the visual display of information."
4. October 11, Garner, "Factors affecting visibility of targets on radar scopes."
5. October 18, Garner, "The reading and interpretation of scales and dials."
6. October 26, Garner, "Speech communication systems."
7. November 1, Gebhard, "The design and operation of controls for human use."
8. November 8, Gebhard, "The layout of panels and equipments."

9. November 12, Gebhard, "The appraisal of equipment from the human engineering point of view."
10. November 22, Chapanis, "The effect of the working environment on human efficiency."

I have always regretted that I never kept at least one hard copy of the illustrations for these lectures. They were works of art as well as instructive.

An amusing incident. Because we discussed military systems, the lectures were classified CONFIDENTIAL, the second highest security level. The first time I came to the lecture hall, I was intercepted by a Marine guard who wanted to see my security clearance. The interchange went something like this:

I: "I don't have anything to show you my security clearance." [I did have a security clearance, but had nothing to show that I was cleared.]
Guard: "I'm sorry, you can't go in."
I: "But I'm giving the lecture!"
Guard: "I'm sorry, Sir, but I can't admit you."

This impasse was resolved only after the guard called the Officer of the Day who conceded that I could enter the hall and give my lecture. I find it paradoxical that trivial happenings such as this one are vividly recalled when so many other more important events are completely forgotten.

Other Publications

Our first book was followed by other influential publications.

Wendell R. Garner, Jack W. Gebhard, Eliot Stellar, Stanley B. Williams and I from Hopkins contributed to a National Research Council book entitled *A Survey Report on Human Factors in Undersea Warfare* (Panel on Psychology and Physiology, 1949). A year later Garner, Gebhard and I from Hopkins were three of the ten persons who jointly wrote another document called *Human Engineering for an Effective Air-Navigation and Traffic Control System* (Fitts et al., 1951). That book was a pioneering effort to formulate a long-range integrated plan for human engineering research leading to an improved air-navigation and traffic-control system. Some of its recommendations are still valid today even as research and development continue on our present system.

On the Allocation of Functions

One entire chapter of our 1951 report was concerned with the roles human operators might assume in future systems. We postulated four possible systems varying in the amount of their automaticity, and we tried to summarize what was known about some general characteristics of human performance, e.g., alertness, overloading, fallibility, sensory functions, perceptual capacities, as well as some things that machines were capable of. In the summary of that chapter we listed some generalizations about how we thought *[H]umans appear to surpass present-day machines and those functions in which present-day machines appear to surpass humans.* [I have italicized our exact words.] We never intended this list (now generally called the *Fitts' list*) to be a prescription for the allocation of functions between humans and machines. In our extended discussion we were careful to emphasize that our hypotheses needed to be tested and validated, to point out the difficulties of conducting research on these matters, and to suggest some of the other considerations, e.g., cost, technical feasibility, that are relevant. Finally, in presenting the list we hedged by saying that we could not foresee what machines could be built to do in the future. Despite our cautions, this list has been published over and over again and it still appears in some current publications as a blueprint for the allocation of system functions. It even figured prominently in a debate held during the 41st (1997) annual meeting of the Human Factors and Ergonomics Society.

The continual misapplication of this part of our 1951 report motivated me several years later to try to set the record straight (Chapanis, 1965c). In that paper I presented my own modified list of some of the relative advantages of men and machines and made this comment about it: "...when we come face-to-face with the practical realities of assigning functions in a genuine man-machine system, we find such general statements of almost no help at all" (p. 3). I then discussed the reasons why I thought this was the case and suggested a strategy for dealing with the allocation problem.

Viewed from my vantage point today, my paper shows its age and, were I to write it today, I would change a number of things. Nonetheless, it was and still is, I believe, basically sound. That other people seemed to think so, too, is evidenced by its having been reprinted in three books of readings (Fleishman, 1967; Schultz, 1970a; Yuki & Wexley, 1971).

A brief digression. I cannot recall my allocation paper without smiling. I read the paper at a symposium on "The design of man-machine systems" held at the XVth International Congress of Applied Psychology in Ljubljana, Yugoslavia, on August 5th, 1964. At the time I was still President of the Human Factors Society and the Society had not yet joined the International Ergonomics

Association (IEA). Etienne Grandjean of Switzerland, Secretary of the IEA, was also in attendance at these meetings, and he contacted me with the suggestion that we meet for breakfast the following morning to talk about having HFS join. I never made that breakfast meeting. I had a room in the Hotel Lvov, the newest in Ljubljana, and on my way down in the morning the elevator malfunctioned. As I recall, it was a full three quarters of an hour before those of us trapped inside could be released. I did meet with Grandjean later in the day, however, and as a result of that meeting I was able to report to the members of the HFS that Grandjean was forwarding to me a proposal that I intended to present at the next Executive Council meeting (Chapanis, 1964b). The HFS did eventually join the IEA, but not until after some spirited discussions in an Executive Council session.

A Highway Safety Research Program

In April 1949, an Advisory Group on Highway Safety Research, the President's Highway Safety Conference, prepared a 29-page report on "A Highway Safety Research Program." The National Research Council was asked to evaluate that report and on June 14, 1950, convened a Conference on Highway Safety Research. I was one of the persons invited to attend that conference and one of the three conferees designated to prepare a summary report on conclusions and recommendations (Chapanis, Marsh & Viteles, June 1950). Our six-page report was submitted to the National Research Council and never published separately. In reading it today I find it a thoughtful and careful blueprint for the multi-faceted research program that has been carried out in the last 48 years.

Chart Reading Under Red Illumination

Because of my experience in the Air Force, I was selected to chair a conference on "Chart reading under red illumination" at the Naval Medical Research Laboratory, New London, Connecticut, on September 30, 1952. The conference had its origin in a request from the Surveyor General of Canada for advice about charts suitable for use in daylight and at night under red illumination. The 14 conferees included two representatives from the Royal Canadian Air Force, three from the US Navy, three from the US Air Force, one from the National Research Council of Canada, two from Johns Hopkins University, and one each from Tufts and Harvard Universities. My report summarizing the results of the conference with some additions of my own was later published by the Armed Forces – National Research Council Vision Committee Secretariat (Chapanis, January 1953).

A Study of Visual Reconnaissance

In 1952 the University received a separate contract from the Air Research and Development Command of the US Air Force for a study of visual reconnaissance from military aircraft. Over a period of about six months nine specialists (S. Q. Duntley, University of California; Conrad Mueller, Columbia University; Lorrin Riggs, Brown University; Robert B. Sleight, Applied Psychology Corporation; S. S. Stevens, Harvard University; and Ward Edwards, H. K. Hartline, M. G. Larrabee, and I, Johns Hopkins University) visited military and civilian agencies that were concerned with this problem, studied relevant reports and documents, and conferred several times about our findings and conclusions. Although the cover page of the 42-page report (Human Factors Scientific Panel, March 1953) does not say so, the Preface identifies the authors of it as Chapanis, Hartline and Larrabee of Johns Hopkins and Captain M. R. De Lucchi of the US Air Force. The report was classified CONFIDENTIAL, the second highest security level, and, although it has undoubtedly been declassified since then, I have never received confirmation of that.

My Book on Research Techniques

My experiences at Bell Laboratories (Chapter 4) motivated me to write a book on research techniques in human engineering (Chapanis, 1959), the first of its kind. It was drafted and largely written under the systems research contract.

CONSULTING ACTIVITIES

We, and I use the collective pronoun to include the entire staff of Systems Research, did not confine our activities to the printed word. Although we did not realize it at the time, our other activities – lecturing to professional groups, providing consulting services, and playing host to visitors – were also instrumental in shaping the field of human factors as we know it today. In a number of cases these activities stimulated other organizations to establish human factors programs of their own.

Progress Report Number 18, dated February 1, 1950, was the first to start listing special activities, such as consulting, lectures, and visits to the laboratory. As I now read through Progress Reports for the succeeding eight and a fraction years[7] I am both impressed and chagrined. I am impressed by the amount of these activities that we collectively, and I individually, engaged in during that period. But I am chagrined to find that almost without exception I have not even a glimmer of recognition for most of the activities these records document.

For example, on January 30-31, 1950, Progress Report Number 19, dated 1 May 1950, states that I was a consultant to the Advisory Panel on Psychophysiology for the Office of Naval Research. There, according to the records, I helped review and advise the Panel about projects in psychophysiology. Later that year Robert B. Sleight and I conferred with engineers from the Glenn L. Martin Company about fire control systems, and in May 14, 1951, I visited the Navy's Special Devices Center, Port Washington, to consult on project "Overhead." Progress Report Number 26, dated February 1, 1952, states that Wendell R. Garner, Robert B. Sleight and I later attended a two-day conference at the Special Devices Center on November 7 and 8, 1951, to discuss a report that had been prepared on the "Overhead" project. I have no copy of the report, and I surmise that "Overhead" was a highly classified project of some sort. What I find disturbing is that I have no recollection whatsoever of any of these activities.

Nor can I recall that in 1951 and in the years following I was consulted by Messrs. G. William Ehreke, Westinghouse Air Arm Division, Baltimore, and John Thorne, Westinghouse Special Products Development Division, Pittsburgh, about tracking behavior and visual displays for new gunsights under development; by Mr. D. T. Hamilton, Cornell Aeronautical Laboratories, about

[7] The last Progress Report, Number 43, is dated November 1, 1958, after which the Systems Research contract ended.

the display of information from computers; by Mr. Carl Moyer, Westinghouse Electric Corporation, about the visibility of targets on CRT faces; again by Mr. Carl Moyer, Westinghouse Electric Corporation, about basic visual data for a display problem; by Mr. J. G. Fleming of the Bristol Company, Waterbury, Connecticut, about the design of industrial recording and measuring instruments; by Mrs. William Shoop of the Training Publications Division, St. Louis Air Force Base, about the use of color in Air Force training manuals; by Mr. Robert Howell of the Westinghouse Corporation about CRT visibility under ambient light conditions at high altitudes; by Mr. R. M. Woodham of the Daniel and Florence Guggenheim Aviation Safety Center, Cornell University, about problems of aviation safety; by Mr. Walter Hancock of the Waverly Press about colorimetric procedures and the color sensitivity of the eye as it relates to the specification of color tolerances for inks and paper in the printing industry; by Drs. Victor McCusick and Samuel Talbot of The Johns Hopkins Hospital about methods of analyzing sound spectra in connection with a proposed research program on heart sounds; and by Drs. John Karlin and Robert Reisz of the Bell Telephone Laboratories about the formation of a new human engineering laboratory.

To continue with my list of forgotten activities, I was consulted by Mr. Jack Kraft, Lockheed Aircraft Corporation, Marietta, Georgia, about instrument panel design; by Dr. John H. Taylor, Visibility Laboratory, Scripps Institute of Oceanography, La Jolla, California, about classified problems in vision and visibility; by Drs. George M. Peterson and Roger J. Weldon, University of New Mexico, and Mr. H. L. Williams, Sandia Corporation, Albuquerque, New Mexico, about human engineering problems related to their work at the Sandia Corporation; by Messrs. W. R. Bonwit, R.

Wagner and K. P. Dolan, Engineering Services Division, E. I. duPont de Nemours Company, about perceptual problems in connection with a work simplification program; by Messrs. D. D. Williams, Engineering Department, Baltimore Gas and Electric Company, and J. Arthur Frantz, Jr., Duralith Corporation, about visual displays; by Dr. H. H. Watts, Martin Corporation, Orlando, Florida, about certain classified military systems; and Messrs. H. H. Rhodes, Aircraft Armaments, Inc., R. S. Grubmeyer and Roger S. Miller, Franklin Institute, and G. M. Wolfe, Jack W. Grewell and G. W. Graham, Airways Modernization Board, about the design of large-scale experiments for evaluating new air traffic control systems.

TALKS

Progress Reports also show that from 1947 to November 1, 1958, when the last Progress Report was prepared, I gave talks and/or chaired sessions in virtually every meeting of the American Psychological Association, Eastern Psychological Association, Society of Experimental Psychologists, and the Armed Forces – National Research Council Vision Committee. Since the exact nature of my participation in these activities can be recovered from appropriate archives, I have made no attempt to itemize them here.

The Progress Reports also show that I gave numerous talks to other groups, among them the Chesapeake Amateur Radio Club, Towson, Maryland; the Industrial Management Club of Richmond, Virginia; the Engineer's Club, Baltimore, Maryland; the Department of Industrial Engineering, Columbia University, New York City; the University of California, San Diego, California; the University of California, Berkeley, California; San Diego State

College, San Diego, California; Naval Reserve BuShips Company 11-16, San Diego, California; Enoch Pratt Free Library, Baltimore, Maryland; Baltimore Section of the Institute of Aeronautical Engineers, Baltimore, Maryland; Electrical Equipment Committee of the Pennsylvania Electric Association, Newark, New Jersey; Connecticut Light and Power Company, Waterbury, Connecticut; Bendix Radio Division of the Bendix Aircraft Corporation, Towson, Maryland; Naval Electronics Laboratory, San Diego, California; Quartermaster Research and Engineering Command, Natick, Massachusetts; Manufacturing Orientation Seminar on Human Engineering sponsored by the American Management Association, New York City; Air Force Office of Scientific Research, Washington, D.C.; Army Chemical Center, Edgewood Arsenal, Maryland; Department of Psychology, Rochester University; Maryland Section of the American Institute of Electrical Engineers; McGill University, Montreal, Canada; and the Rochester Products Division, General Motors Corporation, Rochester, New York.

I have no copies of any of the talks I gave to these groups and I am distressed to have to admit in writing that my memory is completely blank about all of them.

VISITORS

Our laboratories saw an almost constant stream of visitors, both US and foreign, who came to see our facilities and learn about our program and our research. In what follows I enumerate only those foreign visitors who specifically came to see me. For some inexplicable reason, perhaps because I met many of them again from time to time at international meetings or in my foreign travels, I do remember two groups of those visitors. The first, on November

11, 1952, was called the French Applied and Industrial Psychology Study Group and consisted of Drs. Paul Fraisse, Jean Bonnaire, Jean Faverge, Helen Gavini, Suzanne Pacaud, and Pierre Rennes, accompanied by a member of the French Embassy in Washington to translate for us. These individuals were all directors of either institutes devoted to research in applied psychology or psychology departments within industry.

The second delegation came on September 11, 1956, and was sponsored by the European Productivity Agency of the International Cooperation Administration. Members of this group were:

Angelo Iannaccone
Assistant, Institute of Industrial Medicine
University of Florence, Italy

Paul Franz Blau
Director, Personnel and Social Welfare Division
Ministry of Transport and Nationalized Industries
Vienna, Austria

Frederik Hendrik Bonjer
Professor and Chief, Department of Occupational Medicine
Netherlands Institute for Preventive Medicine
Leiden, Netherlands

Kjell Richard Karlsson
Methods Engineer, Institute for Technical Research on Production
Oslo, Norway

Bernard Georges Metz
Professor and Director, Center for Studies of Applied Work
  Physiology
Faculty of Medicine
University of Strasbourg, France

Kenneth Hywel Murrell, Director
Unit for Research on Employment of Older Workers
University of Bristol, England

Friedrich Schofel
Consulting Engineer, Industrial Consulting Management
  Engineering Company
Vienna, Austria

Bernhard Hermann Schulte
Methods Engineer, Institute for Work Physiology
Erlangen, Germany

William Thomas Singleton
Director, Ergonomics Section
The British Boot, Shoe & Allied Trades Research Association
Kettering, England

They were accompanied by Professor Harwood S. Belding, Professor of Environmental Physiology, University of Pittsburgh, Graduate School of Public Health, Pittsburgh, Pennsylvania, and James V. Foley, Project Manager, Office of the Assistant Director for Training and Technical Aids, International Cooperation Administration, Washington, D. C. (As an aside, Belding taught me anatomy and physiology when I was an undergraduate and he an assistant professor at Connecticut State College [now the University of Connecticut].)

In the years following I maintained close contact with Bonjer, Metz, Murrell and Singleton through correspondence and visits.

Having written that, I have to admit to another total memory lapse for a visit on July 25, 1952, from Drs. Marcel Colin, François Canac and Yves Le Grand, and Messrs. Jacques Martin and Roger Raymond. According to Progress Report Number 28, dated August 1, 1952, these gentlemen were part of the French Physiological and Psychological Conditions of Work Team sent to the United States to learn about human engineering under the auspices of the Federal Governments Mutual Security Agency.

My mind is equally blank about visits from Sir Frederick Bartlett, University of Cambridge, England, on April 20, 1953; Dr. W. F. Floyd, Middlesex Hospital Medical School, London, England, on October 6, 1953; Dr. M. A. Bouman, NDRC, The Netherlands, on October 23, 1953; Dr. J. C. Lane, Director of Aviation Medicine, Department of Civil Aviation, Melbourne, Australia, on June 3, 1955; Dr. Horst Mittelstaedt, Max Planck Institut für Verhaltens-Physiologie, Wilhelmshaven, Germany, on October 5, 1956; Professor Koji Sato, Kyoto University, Japan, February 1956; Mr. André Lucas, Renault Motors, France, on February 20, 1957; Major N. Arne 0. Astrøm, Swedish Naval and Aeromedical Research Institution, Stockholm, and Mr. K. Lennart Nordstrom, Systems Analysis Department, SAAB Aircraft Company, Linkøping, Sweden, on June 4, 1958; Mr. George W. Harris, Technical College, University of Birmingham, England, and Mr. Carlo Morgani, Italian Productivity Agency on June 23, 1958; and Professor Rintaro Muramatso, Professor of Industrial Engineering, Wasada

3. Systems Research – 1946-1958

University, Tokyo, Japan, on October 3, 1958.

SYSTEMS RESEARCH AND THE DEPARTMENT OF PSYCHOLOGY

President Isaiah Bowman had almost completely abolished psychology at Johns Hopkins in 1941. To be sure, G. Wilson Shaffer was still a member of the faculty, both as Dean of the College of Arts and Sciences and professor of physical education. John M. Stephens, an educational psychologist, was also on the faculty of the School of Education, but the Hopkins catalog for 1941 (Johns Hopkins University Circular, March, 1941) lists only one course, Introductory Psychology, taught by Sidney Newhall who left shortly thereafter to go to the Eastman Kodak Company. The catalog for 1942 (Johns Hopkins University Circular, March, 1942) lists only one offering in psychology, Introductory Psychology, but identifies no instructor for the course. Clifford T. Morgan is listed in the 1943 catalog (Johns Hopkins University Circular, March, 1943) for the first time which now shows two course offerings in psychology. The next two catalogs, however, show Morgan on leave (to serve with the National Defense Research Committee).

Psychology was not really reestablished as a department until 1946 when Clifford T. Morgan, the newly appointed head, set about to revive it. The systems research project that Morgan brought with him played an indispensable role in the renaissance of the department in the years that followed[8]. Of the 32 students who

---

[8] I also discuss the early history of Systems Research and its impact on the Department of Psychology at Hopkins in an invited address I gave at the G. Stanley Hall Centennial Conference held at The Johns Hopkins University on October 12-13, 1983 (Chapanis, 1986).

received doctorates in psychology from 1947 through 1958 (there were none from 1942 through 1946), 21 had worked as research assistants on the project, and, of the 29 faculty members on the department staff from 1946 through 1958, 22 (or more than 75 percent) were associated with the project in one way or another and received financial support from it wholly or in part.

The systems research project also provided a focus for the department's instructional program. A course entitled "Research in Applied Psychophysics" appeared in the catalog for 1946-1947. It was concerned with the application of psychophysical research to instrument and machine design and was, I believe, the first such course to be taught in any university in the country. In the next year and in the years up to 1958, the psychology department offered courses in industrial psychology, applied psychology, applied experimental psychology, engineering psychology, personnel psychology, and human factors engineering – a varied set of offerings that quickly earned for Johns Hopkins a worldwide reputation for having one of the best (if not the best) programs in engineering psychology – a reputation that lasted until my retirement from teaching at Johns Hopkins in June 1982, when the program was abandoned. For at least a dozen years after that I kept getting inquiries from students who wanted to enroll in my program. One of the great disappointments of my career was the realization that the department had no interest whatsoever in continuing the program that I had been so influential in building up.

THE IMPACT OF SYSTEMS RESEARCH

In reflecting on it now, I think I can honestly say that the research, publications, consulting, lectures, talks, and visits that we all engaged in under the systems research project played a dominant and distinguished role in the advancement of human factors in the United States and probably in the world as well. I am proud to have been part of it.

AFTER SYSTEMS RESEARCH

The termination of Contract N5-ori-166, Task Order I, the Systems Research contract, on October 31, 1958, was followed on November 1, 1958, by Contract Nonr-248(55) between the Office of Naval Research and The Johns Hopkins University, with me as director. Much smaller in size, the new contract replaced the systems research emphasis on radar and Naval information systems with more general problems. Progress reports on the new contract appeared less frequently and were correspondingly briefer. They contain fewer citations of activities such as consulting, participation in meetings, visits, and talks which, from the standpoint of this narrative, makes it much harder for me to document exactly what I did in the years following. Still I persevere.

4. A YEAR AT THE BELL LABORATORIES

1953-1954

In 1953 I was enticed to take a position as Member of the Technical Staff in the Bell Laboratories, Murray Hill, New Jersey. It turned out to have been an unsatisfying move. Accustomed to planning and executing my own research, I chaffed at the necessity to have all my plans reviewed by my supervisor, by his supervisor, and by all the other members of our department. I learned later that this is a common practice in industrial settings and I now understand the rationale for it, but, being fresh from the unfettered atmosphere of a university, I found the practice stifling and somewhat demeaning, as though my supervisors had no confidence in my research abilities.

After having been there a while I began to resent the restrictions imposed on my extramural activities, such as a consulting editorship for the Journal of Applied Psychology and consulting for the Joint Services Human Engineering Guide to Equipment Design. Then, too, having become accustomed to a personal secretary to whom I could hand material with specific instructions about how I wanted things typed, I now had to turn material over to the head of a typing pool who had the material typed by some anonymous typist with whom I could not communicate. Having some idiosyncratic ideas about how I wanted my work done, I was often dissatisfied with the way material was returned to me. In desperation I pleaded for and finally was allowed to have my own typewriter to give me the greater freedom I felt I needed.

During all the time I was at Bell Labs I could never completely sever my ties to Hopkins, and on most weekends I would either drive or take a train back to Baltimore so that I could chat comfortably with students and staff and continue working on unfinished research projects there. Throughout the year the strength of those ties increased as did my disaffection with Bell Labs.

THE RESEARCH CLIMATE

The laboratories had a large number of distinguished physical scientists and engineers who were doing some outstanding work. Research on the psychophysics of hearing, for example, by such men as Fletcher and Riesz, was excellent, but the prevailing attitudes towards behavioral research, although not hostile, was certainly not sympathetic. I remember discussions with John Pierce, an eminent physicist, who found it difficult to understand my emphasis on careful experimental design and the importance of controls in doing behavioral studies. The prevailing unspoken sentiment seemed to be that it was all more or less common sense and since they were human they knew how telephones, codes, telephone booths, and telephone directories should be designed.

I had encountered attitudes of this kind in the Air Force when at a mock-up inspection for a new aircraft, a high-ranking officer with years of flying experience would get into the cockpit of the mocked-up aircraft and on the basis of inspection alone pronounce it satisfactory or requiring redesign. I had not, however, expected to find such attitudes at Bell Labs. After all, these were highly educated men, scientists themselves. It was a dramatic change from the university where behavioral research was not only accepted, but encouraged.

SOME PERSONALITY CLASHES

It was even more disturbing to find that my immediate supervisor, John Karlin, a psychologist with a background in the psychophysics of audition, did not understand how to design and conduct properly controlled experiments. He had come to Bell Labs from S. S. Stevens' Psychoacoustic Laboratory at Harvard where multi-factor studies were not the norm. As a result, I had serious, sometimes heated, arguments with him about how to do behavioral research.

A Poorly Designed Study

One of Karlin's studies, for example, was intended to discover whether tilting a telephone keyset would improve keying performance. Two girls keyed out numbers one hour a day for eight days with a keyset that lay flat on the operating board. On the ninth day the keyset was tilted to 20° and the girls used the keyset in that position for the next three days. The results Karlin reported were that, on the average, operators had a keying time of 9.0 seconds for the eight days of test at the 0° tilt, and 7.7 seconds for the three days at the 20° tilt. When I plotted mean times against day of test, however, I found that the results could very well have been explained by continued learning throughout the entire series of trials. In my book on research methods (Chapanis, 1959) I used this study as an example of how not to design experiments. I'm sure none of my graduate students would have designed an experiment as naive as this. My plot of his data appear as Figure 26, page 155, in my book.

A Replication of Karlin's Study

I then designed a study to examine the same question that Karlin had attempted to answer. My study, however, was designed as an 8 X 8 Latin square with the three principal variables being tilt (0° to 40°), day of test, and subjects. The results of that study (Scales and Chapanis, 1954) appear as Figure 29, page 194, in my book on research methods. Tilt of the keyset had no demonstrable effect on performance. As expected there was a significant and marked improvement in performance on successive days, and there were, of course, large differences among subjects.

STUDIES OF TEN-BUTTON KEYSETS

The only other substantive study I was able to complete during my tenure at Bell Labs was motivated by a practical problem. It's hard to remember that in 1953 there were no push-button telephones, only rotary dial telephones. The telephone company was planning to go to push-button technology, but had only one model of a push-button keyset. That was on the toll-operator's station. The numerical part of the toll-operator's keyset was two vertical rows of five keys labeled this way:

4	5
3	6
2	7
1	8
0	9

Some Error Data

Another reminder of the state of technology in those days: subscribers could not dial long distance numbers directly. They had, instead, to call a toll-operator, tell the operator the number they wanted, and the toll-operator made the connection by keying it on her keyset. Although it was (and is) illegal for anyone to listen to long distance telephone calls, it was possible to do what was called "service observing." Numbers requested by subscribers could be recorded and compared with numbers that were keyed by long-distance operators. Service observations of thousands of telephone calls in Washington, D.C., showed, as I recall, that about 13% of numbers were incorrectly keyed by experienced operators. I spent a lot of time studying these data looking for patterns that might account for the errors. I could find only one explanation that seemed to account for some of the errors, namely that the operator expected numbers to be someplace else. The question that occurred to me was,

4. A Year at Bell Laboratories – 1953-1954

"Where do people expect to find numbers on keysets?"

Expected Locations of Numbers on Keysets

The study I designed (Lutz & Chapanis, 1955) was, from my perspective today, a good one. We tested six different keyset configurations – one with two vertical rows of five blank keys, one with two horizontal rows of five blank keys, and four 3 X 3 arrangements of blank keys with the 10th blank key above, to the right, below and to the left of the square. Presentation of the keysets was counterbalanced, numbers were presented in a random order, and subjects were selected according to a carefully designed stratified plan. Briefly, the results showed that most preferred pattern of numbering for all keyset configurations was to have numbers increase from left to right and from top to bottom. Of the six arrangements the one on which there was the greatest agreement was the 3 X 3 square arrangement with the 10th key below the square – the arrangement you now find on all push-button telephones. This, so far as I know, was the first study to have been made on this problem.

Performance Tests of Ten-Button Keysets

I left Bell Labs before doing the next logical study, namely, to find out if keying times and errors were different for the different keysets. A study of that type was performed by Deininger (1960) but in my critique of it (Chapanis, 1963a, p. 312) I pointed out that the study was flawed with so many confounded variables one could not legitimately draw any positive conclusions from it. In a report of his work, however, Hanson (1983) said that

Deininger's studies led to the selection of the now ubiquitous *Touch-tone* telephone dialing arrangement. His studies showed that it was strongly preferred over the now equally familiar calculator arrangement (p. 1578).

That was a misstatement of the facts. Deininger's study (1960) did *not* show that the *Touch-tone* dialing arrangement was strongly preferred over the calculator arrangement. He used very small numbers of subjects and made no attempt to evaluate the statistical significance of the results he obtained. My analysis of his preference data showed that many of his distributions of votes did not differ significantly from chance (Chapanis, 1963a, pp. 311-312). Several years later Conrad and Hull (1968) designed a more definitive study and concluded that the telephone layout had "a small speed advantage and a highly significant accuracy" (p. 170) advantage over the adding machine layout.

MY DEPARTURE FROM BELL LABS

Because of my strong opinions, which I never hesitated to express, I left Bell Labs under a cloud. Years later Hanson (1983) published an article and a bibliography of human factors research that had been conducted at Bell Labs and conspicuously omitted citing my two papers even though both were attributed to Bell Labs.

A Traumatic Leave-Taking

For the fourth of July holiday while still technically in Bell Lab's employ, my family and I visited my parents in Connecticut, and on July 3, 1954, my wife and I went riding. I was thrown from my horse against a tree, breaking 10 ribs and collapsing a lung. I arrived at the Danbury, Connecticut, Hospital more dead than alive, and for the next week lay flat on my back in an oxygen tent while I clung tenuously and agonizingly to life. After an additional four weeks my body had healed itself enough so that I could be discharged from the hospital encased in an enormous bandage that encircled my torso from waist level to shoulder height. By mid September I had recovered sufficiently to take on a light teaching load at Hopkins, to which I had come back gratefully. I never returned to Bell Labs.

5. THE ORIGINAL
HUMAN ENGINEERING GUIDE
TO EQUIPMENT DESIGN
1953-1963

From 1953 to 1963 a considerable amount of my time and energies was consumed in the preparation of a *Human Engineering Guide to Equipment Design* (Morgan et al., 1963). I quote several paragraphs from the Preface to that work and then add some supplementary notes.

"In May of 1952, the Panel on Human Engineering, Committee on Human Resources, Research and Development Board, Department of Defense recommended that the three branches of the armed services jointly develop the "Human Engineering Guide to Equipment Design." A memorandum to the Secretaries of the Departments of the Army, Navy and Air Force submitted by the Research and Development Board in October of the same year recommended that both design-engineering and human-engineering specialists be included in a Joint Services Steering Committee for the Guide.

"The three services responded to this memorandum by appointing members to the steering committee and by providing the necessary funds. The Navy accepted responsibility for coordination among the services, and the Secretary of the Navy directed the Chief of Naval Research to assume this responsibility. The Chief of Naval Research, in turn, assigned responsibility for coordination to Dr. Henry A. Imus of the Office of Naval Research.

"The first meeting of the Joint Services Steering Committee was held in February of 1953, and Dr. Arnold M. Small of the U.S. Navy Electronics Laboratory was elected chairman and authorized to appoint an executive council. On March 31 and April 1, 1953, the executive council, which consisted of Drs. Walter F. Grether (Air Force), Franklin V. Taylor (Navy), Lynn E. Baker (Army), Clifford P. Seitz (Navy), and Alphonse Chapanis (Air Force) in addition to Drs. Imus and Small, met to decide, among other things, the purposes to be served by the Guide.

"By December of 1953, a detailed plan for the Guide had been drawn up, and, by March of 1954, work had begun on all but four of the proposed chapters. In August of 1955, a style manual to assist in preparing the Guide for publication was prepared by an editorial working group consisting of Drs. Small, Grether, and Chapanis and Wesley E. Woodson.

Later, Drs. Imus and Small left the Navy, and Dr. Max W. Lund of the Office of Naval Research was assigned responsibility for chairing the group and coordinating its work. In June of 1956, Dr. Clifford T. Morgan joined the group as consultant and editor. In May of (p. vii) 1960, Jesse S. Cook III was assigned to work with the group in the actual preparation of the Guide for publication" (p. viii).

PREPARATION OF THE GUIDE

The preparation of the Guide was beset with many difficulties. The 11 chapters had been assigned to individuals who the steering committee thought were most competent to write them, but contributors were frequently late in submitting their drafts, and some submitted material was judged unsuitable and had to be reassigned or rewritten by other individuals. For example, J. D. Folley and J. W. Altman of the American Institute for Research had submitted some material for Chapter 9, "Design for ease of maintenance," but it had to be rewritten by Cook and me. I also wrote one section of Chapter 5 on "Man-machine dynamics," because the original material did not adequately explain and illustrate machine transfer functions.

In addition to my reading and reviewing all the chapters, they were reviewed by other experts and by representative design engineers. Receiving and sending out chapters and reviews and corresponding with authors and reviewers involved a great deal of paperwork because some chapters went through as many as four iterations. Since there was no way I could handle all this work, I strongly urged that Morgan be appointed effectively editor-in-chief. I admired Morgan greatly for his editorial and managerial abilities. He had published successfully (he was a consulting editor for McGraw-Hill), he and I had worked together on several writing projects, and our offices were close to each other so that we could communicate with a minimum of paperwork. The steering committee agreed and appointed him to that position in June 1956. It was a happy choice.

At one time, to go about our work most effectively, Morgan and I decided to get away from Baltimore and the distractions that inevitably intruded on us. For a couple of months during the summer of 1958 or 1959 (I can't remember which) we, a secretary, typewriter, Dictaphone, and files moved to a motel on the West Coast near San Diego. I don't remember exactly where it was, but it was close to a beach because I had insisted that I needed to take a daily swim. In any case, it was a highly productive summer with us often working late into the night.

When the manuscripts were nearing completion in 1960, McGraw-Hill designated a production editor, Jesse S. Cook III, to prepare it for publication and to oversee it through the production process. Lund was not only technically responsible for chairing the working group and coordinating its work, but also for providing the funds for the project. Although his editorial contributions were minimal, we felt that we could not ignore his request to be listed as one of the editors.

The book was published by a commercial publisher under contract with the U.S. Government. Contrary to some rumors, neither the editors nor the contributors received any royalties. In 1965 the book was published in Japanese by the Corona Publishing Company, Ltd, Tokyo. It had been translated by Professor Takashi Kuroda, at Nihon University. Professor Kuroda was my host when in 1982 I spent six weeks lecturing in Japan. I have made no attempt to include the Japanese title in my bibliography.

TWO BYPRODUCTS

A Book on Methodology

In 1954 or thereabouts I felt that we needed a treatise on methodology in human engineering. The applied experimenter faced problems – choice of subjects, realism, choice of dependent variables – normally not serious considerations in experiments done in academic settings. In addition, methods other than controlled experimentation, for example, link analysis, critical incident studies, and accident investigation, were not normally taught in graduate schools although they could yield important data in applied settings. With that in mind in 1955 I prepared a rather long essay which I titled *The Design and Conduct of Human Engineering Studies*. The eight chapters were titled "Introduction," "Methods of Operational Observation," "Methods for the Study of Accidents and Near Accidents," "The Experimental Methods," "The Psychophysical Methods," "Statistical Methods," "Some Special Problems of Experimenting with People," and "The Last Word." In reading it today I find it a competent piece of work, as far as it went.

I'm not sure now why I wrote it. It is possible that I proposed the project to the Joint Services Steering Committee and they authorized it. It's more likely, though, that they initiated the request that I write such an essay as the first chapter in the Guide. In any case, the title page of the published document states that it was "Prepared for the Joint Services Steering Committee for the HUMAN ENGINEERING GUIDE TO EQUIPMENT DESIGN." In the first chapter I said that "It is written for engineers and other technical specialists who are not primarily psychologists, but who, for one reason or another, are required to do experiments involving men and machines" (p. 3). That notwithstanding, the Steering Committee judged it inappropriate for inclusion in the Guide, and rightly so. Instead, they authorized its publication by the San Diego State College Foundation under Project NR 145-075, Contract Nonr-1268(01) between the San Diego State College Foundation and the Office of Naval Research (ONR). The 73-page booklet was published in July, 1956, as Technical Report Number 14 under San Diego's Contract (Chapanis, 1956).

I'm not sure how many booklets were printed or what San Diego did with the booklets. I know that I was given a small quantity (perhaps 200 or so) for my own distribution. I later used this report as the basis for a greatly enlarged book that I published in 1959 (Chapanis, 1959).

A Human Engineering Bibliography

One purpose of the San Diego contract with ONR was "to conduct bibliographic research and collect information concerned with human factors in equipment design." From 1953 to 1956 several thousand bibliographic items had been collected by the San Diego staff. So many requests were received for copies of the bibliography that the Executive Council of the Joint Services Steering Committee decided the bibliography should be published.

In July of 1956 my family and I spent the summer in San Diego in order that I might assist in the publication of the Bibliography. According to the Preface I was involved in three tasks: (1) delimiting the areas to be covered by the Bibliography and establishing criteria for including or excluding any particular item, (2) creating a classification system for the Bibliography, and (3) classifying each item into appropriate categories. I also supplied a number of bibliographic items that I had brought with me from Baltimore and that I had discovered through library work at the college. My criterion for selection was simple to state, but more difficult to use. If, in my opinion, an article or report contained information of practical value to a designer, it was included. If it did not, it was excluded.

The fourth task – completing the work of getting all items into a consistent bibliographic form, preparing the manuscript, and arranging for publication – was assumed by McCollom. The 128-page Bibliography was published in November 1956 as Technical Report Number 15 under San Diego's contract with ONR (McCollom & Chapanis, 1956). It contained 5,666 items classified under 94 headings. Hundreds of items were cross classified if they contained information relevant to more than one category. A three-page Preface identifies everyone who had even a minor role in the compilation of the Bibliography.

Once again, I don't know how many bibliographies were printed or what San Diego did with them. I do know that I was given a reasonable number for my own use and distribution.

SUMMARY

In retrospect, I think I can honestly say that these publications were major accomplishments and I am proud that I was able to have had a significant role in their preparation. The *Guide* for the first time systematized and organized virtually all human factors knowledge for the benefit of researchers and equipment designers.

5. The Original Human Engineering Guide – 1953-1963

To be sure, Woodson (1954) had already written and published the first of what, in succeeding years, were to become a series of extremely popular and successful Guides. But without in any way disparaging Woodson's book, it had neither the depth nor the sophistication of ours. Substantial portions of our Guide were incorporated without change in the revised version that was published nine years later (Van Cott & Kinkade, 1972). Of the two lesser publications, my book on methodology was also an important forerunner of others that were to follow.

6. RESEARCH ACTIVITIES
1958-1985

When the Systems Research contract ended on October 31, 1958, it was followed immediately on November 1, 1958, by Contract Nonr-248(55) between the Office of Naval Research and The Johns Hopkins University with me as director. That contract was followed by Contract Nonr-4010(03) on May 16, 1963, between the Office of Naval Research and the Department of Psychology, The Johns Hopkins University, and it, in turn, was terminated on 14 July 1970. All the publications dated between 1958 and 1970 (or 1971 because of publication lags) that I have cited elsewhere in this book were prepared under one or the other of these two Navy contracts. The year 1970, however, saw a major change in the focus of my research.

COMMUNICATIONS RESEARCH

Starting in 1970 and continuing for the next ten years I had research support from the National Aeronautics and Space Administration (Research Grant NGR-21-001-073); National Science Foundation, Office of Science Information Service (Research Grants GN-890 and GN-35023); Office of Naval Research (Research Contract N00014-75-C-0131); and National Science Foundation, Division of Advanced Productivity Research and Technology (Research Grant APR-7518622) for a program on interactive communication. From my records I am able to confirm the inclusive dates for only the ONR contract. It extended from October 1, 1974, to June 30, 1980.

Goals of the Program

The goals of the program were to discover (1) how people naturally communicate with each other when they are required to do various kinds of tasks, (2) how interactive human communication is affected by the machine devices and systems through which people converse, and (3) what significant system and human variables affect interactive communication.

The Research Setting and Laboratory

My laboratory started out with just two rooms, but was expanded in 1974 to four rooms (Figure 4). Four basic channels of communication were tested: video (the picture part of television without the voice), audio, handwriting, and typewriting. The latter three were tested singly. For example, anything written on a telepen device appeared simultaneously on the other three, and anything typed on an input-output writer appeared simultaneously on the other three. In addition, all channels of communication were tested together in various combinations, for example, voice and typewriting; voice, handwriting and typewriting; and so on.

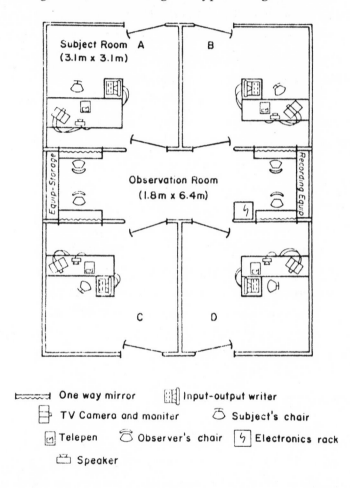

One way mirror   Input-output writer

TV Camera and monitor   Subject's chair

Telepen   Observer's chair   Electronics rack

Speaker

Figure 4. My communication research laboratory in Ames hall, The Johns Hopkins University, 1975-1979.

Problems

For our research we generated a large inventory of tasks or problems. About the only feature they had in common was that they were all of recognizable practical importance. They were real-world, everyday problems about which you might want to consult someone for information, have a conference, or solve by communicating interactively with a highly sophisticated computer.

Research Output

Altogether we (my students and I) produced and wrote 19 articles about our work. The best summaries of our findings are in two articles I wrote (Chapanis, 1979b, 1981). The former was an invited talk I gave at a course organized by the Institut National de Recherche en Informatique et en Automatique held in April 1979 at the Chateau at Gif-sur-Yvette, France; the latter was for an invited address I gave at an Advance Study Institute sponsored by the NATO Advanced Study Institutes Program held at Mati, Attica, Greece, September 5-18, 1976. I also gave an invited address on our research findings on August 31, 1978, at the Eighty-sixth Annual Convention of the American Psychological Association in Toronto, Ontario, Canada.

An Evaluation

Looking back on them now, I feel they were good studies – innovative, well designed, carefully executed and analyzed. I occasionally see some of them cited in current literature. Still, I am disappointed in them because they have had no discernible impact on technology. Although they tested real-world tasks and used sophisticated communication systems, they were essentially basic studies. And that is precisely what was wrong with them. They tested no real system, nor were they designed to attack any genuine problem. They were just basic studies – abstractions.

If I compare them with the little bearing counter studies I made in 1946-1949 (see pp. 39-41), I have to admit that the bearing counter studies were simple and, in a sense, trivial. They were, however, designed to answer a practical question: Can an operator obtain bearings on a PPI faster and with fewer errors with a counter than with a dial? I can't prove that my studies were responsible for bearing counters being incorporated in virtually every radar made since that time, but I think they probably had something to do with it. If so, they are examples of small studies with a large impact.

Or, I can compare the communication studies with the simple study Lutz and I did on where people expect to find numbers on telephone keysets (see pp. 73-76). That study was also done to answer a practical question: What should a push-button telephone keyset look like? Once again, I can't prove it, but I think it's more than coincidence that the telephone company settled on the keyset arrangement we found best in our study and that millions of telephone handsets since then have been configured that way. In any case, I know one thing for sure: Our study has been cited in ANSI/HFS 100-1988, albeit inaccurately (Chapanis, 1988a), as providing support for one of the recommended layouts of numeric keypads for computers. So, once again it's an example of a basically simple study with an impressive impact.

Our communication studies were much more complex than the

two simple ones I've cited, and they discovered a number of things about the way people communicate interactively, but the technology of teleconferencing and human-computer interaction has developed without any recourse to our findings. I think there is an important lesson here for all human factors researchers.

STUDIES OF HUMAN-COMPUTER INTERACTION

Sometime in 1979 the University and I signed a research agreement with IBM's Systems Coordination Division and later with IBM's Corporation Systems Products Division whereby IBM would support work on human-computer interaction provided that we made our findings available to IBM before the findings appeared in print. Not only did IBM agree to support my work, but they also agreed to provide me with computer terminals and a hard-wire connection to IBM's own computers. This was now a greatly expanded effort for which the small research rooms we had in the Department of Psychology would no longer suffice. Accordingly, I moved into two suites of a new commercial building (Suites 302 and 303, 7402 York Road). Since we agreed to move before the building was entirely completed, I was able to have the laboratory configured to my own specifications (Figure 5).

The new laboratory, officially the Communications Research Laboratory (CRL), was the finest I have ever had. It was handsomely furnished with new teak (or in some rooms oak) desks, cabinets, file cases, and book cases. We had a large conference room for meetings. Telephone service was provided through four different lines. Most important, I had sufficient funds to provide the laboratory with virtually anything we needed for our research.

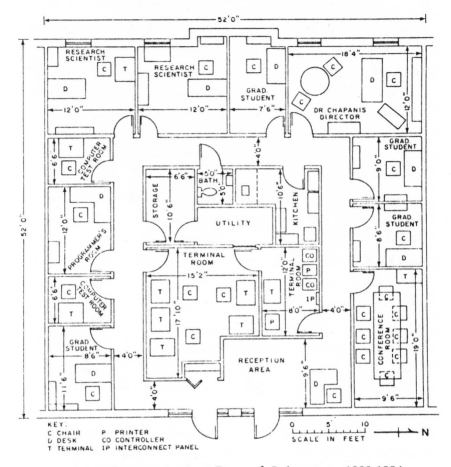

Figure 5. My Communications Research Laboratory, 1980-1984.

The laboratory was staffed with two full-time Ph.D. research associates, W. Randolph Ford and Gerald D. Weeks, both of whom had received their degrees with me. In addition, I had two full-time secretaries, Margaret Iwata and Julia Nardi, one of whom was at the office I still maintained on campus; a part-time computer programmer, Arthur Giovanetti; five graduate students, Janan Al-Awar, John F. Kelley, Kathleen Potosnak (Neumann after her marriage), Joan M. Roemer, and Elizabeth Zoltan (Zoltan-Ford after her marriage); and several undergraduate students, Alex Hauptmann, Spencer Johnson, Daniel Schwartz, and Marc A. Sullivan. Being away from the University had its advantages. We made a close-knit research group and the first two years that I spent there were some of the most satisfying I had ever had as a researcher.

Research Studies

In all I, either alone or with co-authors, published 7 articles stemming from our computer research. I described the overall research plan for and goals of the laboratory in an invited address I gave at a symposium held at the University of Houston in 1980 (Chapanis, 1982a). Two studies are worth special comments. The first study to be published under the program (Al-Awar, Chapanis, & Ford, 1981) received the IEEE Professional Communication Society Outstanding Article Award in 1982. Another early study was a large survey of what professional persons – physicians, accountants, lawyers, and pharmacists – thought about computers (Zoltan & Chapanis, 1982), a study that I see cited from time to time in the current literature.

GTE Supported Studies

In addition to support from IBM I had funds from the GTE Laboratories under the terms of a letter agreement dated November 2, 1980, and terminated on November 2, 1981, for research on "Human factors of voice/data communications." GTE also provided us with a few terminals linked to their mainframe computers in

Waltham, Massachusetts, and a Telemail terminal also linked to their central computers.

Five projects were undertaken in this program: (1) The first was a study of potential applications for voice-enhanced computers in hospital systems. This work was done in one of our local hospitals by two of my graduate students, Kathleen M. Newman and Elizabeth Zoltan. The report was submitted to GTE, but was never published in the periodical literature. (2) Janan Al-Awar, another of my graduate students, worked on an instrument to evaluate the use of speech in person-computer interactions. This work was completed under IBM auspices as her Ph.D. dissertation (see Table 2, p. 107). (3) Kathleen Neuman also did a study on the effect of input verification on voice and keyboard data entry. A report of this work was submitted to GTE but was never published. (4) Elizabeth Zoltan, Gerald D. Weeks and W. Randolph Ford did another unpublished study on a comparison of voice and keyboard inputs on natural language communication with computers. (5) Daniel Schwartz, an undergraduate student, performed an evaluation of GTE Telenet's Telemail system and its Telemail Trainer, both of which he found to suffer from some serious human factors deficiencies.

I have always regretted that this work never resulted in any publications. Some of it was very good. In part, I think it was because we were trying to do too much in too little time. I did, however, use several of the human factors deficiencies Schwartz discovered about GTE's Telemail to illustrate an invited lecture I gave at a Human Factors symposium sponsored by ITT Europe Engineering Support Centre and held at the Excelsior Hotel, Heathrow, London, on May 18 & 19, 1982 (Chapanis, 1982c). These included deficiencies in the design of the hardware (the keyboard

94

and terminal), the terminal display, and the software.

THE DEMISE OF THE COMMUNICATIONS RESEARCH LABORATORY

My luxurious and comfortable existence in the Communications Research Laboratory was, unfortunately, soon to come to a sad and bitter end. March 17, 1982, was my 65th birthday and the University had a firm policy about retiring professors at age 65. In 1981 I pleaded before the entire faculty that I be allowed to continue beyond age 65, but my plea was fruitless.

I Reap Antipathy for What I Had Sown

That unsuccessful plea was followed by a private meeting with Dean Sigmund Suskind and the Chairman of the Department of Psychology, William D. Garvey, during which I offered to support myself from research grants and asked only to be allowed to keep my title of professor and have the privilege of selecting new graduate students to study and receive their degrees with me. Immediately after hearing my proposal, the Dean showed me four letters written at Garvey's request by full professors in the Department of Psychology. All stated in the strongest possible language that I should *not* be allowed to continue past the mandatory retirement age. I was crushed. I left the meeting abruptly because I could scarcely contain the tears that welled up in my eyes.

I had known for a long time that I was a maverick in the Department of Psychology, but my research and my professional activities had been so successful and time-consuming that I scarcely took notice. For one thing, most of the department had become strongly and firmly committed to basic research, and I never hesitated to make my views known about what I thought was the sterility of most basic research. In addition, I had more research funds than any one else in the department and, for a time, more research funds than all the other members of the department combined. In fact, some full professors who lacked research support would occasionally come to my secretary to beg for supplies that they could not afford. Between 1947 and 1982 I trained more graduate students than any other member of the department. I published regularly, traveled widely, consulted often, and received several prestigious honors. All of this created a generous amount of enmity and jealousy.

Departmental feelings manifested themselves in several ways. When I returned in 1961 from my tour of duty at the US Embassy in London no one congratulated me or expressed any interest in what I had done. In the years following, I frequently had foreign visitors and on those occasions would often entertain them in my home to which I invited the entire department. Although most members of the staff came to these parties, my wife and I were never once invited to the home of any of the persons who had enjoyed our hospitality. When I received the prestigious Distinguished Contribution for Applications in Psychology Award from the American Psychological Association in 1978, not a single member of the department congratulated me or even took notice of my award. I recognized all these symptoms in some casual sort of way, but was so engrossed in my own affairs that I never let it concern me greatly. But now in 1981 all that accumulated hostility had come home to me – and with a vengeance!

I was officially retired on June 30, 1982. The irony was that in the following year the government changed the rules to make 70 the mandatory retirement age for college professors.

A Token Assignment

In the early 1970s Dean George Owen had put me in charge of the Board on the Use of Human Subjects in Research, to deal with ethical concerns that were becoming increasingly insistent everywhere in the country. I did such a good job of organizing and administering this activity that Dean Suskind, who succeeded Owen, asked me to continue doing so for another year after my official retirement, that is until June 30, 1983. In return I was allowed to keep my office in the Department of Psychology and was given another part-time secretary to keep the files, schedule Board meetings, take the minutes at the meetings, and handle the correspondence that was associated with this activity. I was, however, severed from all department activities and from acquiring any new graduate students.

I Silently Steal Away

When I finally removed the last of my files, books, and belongings from my office in the Department of Psychology on June 30, 1983, no one came to say "Good bye," shake my hand, or even take any notice of my leaving. I have always felt it a deliberate affront that I was never proposed for the title of Professor Emeritus.

And Take Refuge in my Laboratory

Meanwhile, I still had my sanctuary in the CRL, but that was not to be for very long. On June 30, 1982, the University transferred to me the responsibility for paying the rent on CRL, and for another year and a half I paid the rent from the funds of my own corporation, Alphonse Chapanis, Ph.D., P.A. [about which more later]. IBM accommodatingly allowed me to keep the computer terminals and linkage to its main computers until my last three students (Kelley, Neumann and Al-Awar) had completed their research, but, with the prospect of my having no more graduate students, discontinued financial support. With the loss of funds, I was forced to terminate some staff positions and stop paying most undergraduate students.

In 1982 I continued to support three students (Kelley, Neuman and Al-Awar) until they completed their dissertations. At about that time I hired an undergraduate student, Gordon Rubenfeld, on an hourly basis to support Neumann with her computer programming, and hired Kristine A. Haig as a part-time laboratory assistant. Kelley, Newmann and Zoltan-Ford completed their dissertations in June 1983. Although Al-Awar had completed her computer work in 1983, she did not actually finish writing her dissertations until 1985 after I had closed CRL.

A Brief Reprieve

For a short while in 1982 Edmund Klemmer joined me in securing some funds for some small studies for the Office Products Division of IBM in Raleigh. In addition to supporting Neumann, Haig worked with Klemmer, work that eventually resulted in a publication (Klemmer & Haig, 1988). Haig also scheduled and tested subjects for a small study I designed on hazards associated with signal words and colors on warning signs, a study I did not get around to publishing until years later (Chapanis, 1994). Gradually my students and supporting personnel left until in December 1983 I was left with an empty laboratory shared only with my long-term secretary, Margaret Iwata.

I Open My Consulting Offices

Maintaining a largely empty laboratory was more expensive than I could afford, so on December 30, 1983, I closed CRL and rented a smaller office of only five rooms (Suite 210, Ruxton Towers, 8415 Bellona Lane, Towson, Maryland 21204). I furnished it with whatever desks, book cases, credenzas, tables and other assorted items of equipment I needed from CRL and donated the rest to the Department of Psychology. I also brought my secretary, Mrs. Margaret Iwata, to my new offices.

The move to these new offices meant the end of my research activities. To be sure, I did bring in people from the surrounding area from time to time to evaluate equipment and products for such companies as Owens-Illinois, but my new offices were used primarily for small conferences, consulting, and writing.

THE AFTERMATH

I missed doing research. I missed the intellectual challenge of designing a study, the excitement of seeing data come in and of analyzing the data, and the thrill that comes when results fall into place to tell me something I had not known before. On the other hand, writing the results up and publishing them have always been a chore. I knew what the results were; putting them into a manuscript was a kind of anti-climax. And that may explain why I still have sets of data – good data – in boxes that I keep promising myself I *will* publish. Some day.

7. MY STUDENTS
1949-1985

In some ways I think the greatest legacy I have left the human factors profession are the students I trained. Actually, *trained* may be too strong a word because it implies a direct transfer of knowledge from me to them. It is probably more correct to say that I merely provided them with an environment in which they could train themselves. In any case, the greatest pleasure I derived from my teaching career at Hopkins was from my students. They were a hard-working, bright, imaginative group and really fun to work with.

One incident may perhaps convey some idea of the cordial relationship we had with one another. I was lecturing about some problems for computers designed to be usable in diverse cultures. I said something like, "For example, computers have to be programmed to print backwards if they are to deal with languages like Arabic." At that point, Janan Al-Awar, a Lebanese student, raised her hand and said, "No Professor, you read Arabic normally. It's English that you have to read backwards." Needless to say, the other students in the class and I cracked up and could hardly stop laughing.

Virtually all my students have gone on to successful careers in universities, industries, and government. Yet I recall how often, especially after the termination of Systems Research, I had to argue in staff meetings about admitting most of them to graduate study. From about 1970 onwards, members of the psychology department felt that my research was too applied, not theoretical enough. The prevailing sentiment was that graduates of our department could be said to have achieved success only if they became teachers at some college or, preferably, some university. I, on the other hand, argued that a graduate who becomes a manager or vice president in some firm, or is advanced to high rank in some branch of the armed services, or becomes the head of some branch of the government is fully as successful as one who becomes a professor. I was and still am proud of my students – all of them. It pleases me that some of the students for whom I had to fight most strongly have achieved greater success in my eyes than many other students who were readily admitted to other programs in our department and have since vanished in the mists of mediocrity.

MY TEACHING PRACTICES

One of my favorite courses was the analysis of variance and design of experiments. In the first lecture I told the class something like this: "If you ever have a question, I want you to feel free to ask it at any time. I will never laugh at or ridicule your question, but will try to answer it as best I can. Keep in mind, if you have a question, there are probably some other students who have the same question, and that means that I am at fault for not having explained something clearly enough."

Getting Involved in Research

I also liked to have my students become actively involved in research from the day they joined my program. At first, they were apprentices – helping me or some senior student on research in progress. As they advanced they became partners in research contributing their ideas to research underway. I held weekly research seminars at which each student discussed what he or she was doing or planning to do, talked about any problems or difficulties encountered, and sought help if needed. As a result, each student then moved confidently and easily into his or her thesis research when it became time to do so.

Writing Style

I was always a strong believer in writing simply and clearly. When a student submitted a piece of writing to me, I insisted that it be triple-spaced to allow me ample room to make corrections and suggestions. I criticized style of writing as much as content, because I said that the best science in the world is of no use if people can't read and understand what you did. I also had the students make successive drafts, as many as four, of their theses until I was satisfied with them.

This was not a one-way street. The draft of every speech, every article I wrote was also typed triple-spaced. Copies were distributed to all my students with the following instructions: Feel free to criticize anything – logic, content, style of writing. Don't hesitate because I happen to be the author. But anytime you make a criticism, tell me what I should have said or done. Don't just tell me what's wrong; tell me how to correct it.

I have been complimented from time to time on the clarity and precision of my writing. To the extent that is true it is because of the insightful comments and criticisms I received from my students. It was a learning experience for both them and me. One of the things I missed most about leaving the university was that I no longer had a sounding board to tell me when I was being vague, illogical, or just plain obscure.

SOME ACCOLADES

Of all the compliments I have received for the speeches I have given or the things I have written, the ones that touch me the most, that warm my heart, and that I remember longest are the acknowledgements my former students have made in their Ph.D. dissertations. They are sincere and honest expressions of appreciation and reminders of the close working relationships we enjoyed over the years. Here are a few:

> And to my advisor, Professor Alphonse Chapanis, I wish to express my deepest appreciation for his guidance, support, and patience throughout this long endeavor. I am honored to be one of his students and am pleased to call him friend. I trust I have incorporated some of his unyielding demands for excellence, precision, and accuracy in the field of research. (Brecht, 1979, p. ii)

7. My Students – 1949-1985

Words, words, words cannot describe my indebtedness to professor Alphonse Chapanis. He is known around the world for his achievements in the field of human factors engineering, and around the lab for his enthusiastic willingness to share his knowledge, experience, and insight. He treats his students with respect, is always supportive of them, and I for one can attest to his saintly patience. He truly provided me with a unique educational experience. I am proud to have been his student, and hope that one day he'll be proud of me. (Roemer, 1981, p. V)

My advisor, professor Alphonse Chapanis, also provided financial support. More importantly, however, he was responsible for the guidance and training I received at his Communications Research Laboratory during the past four years. I am proud and grateful to have been one of his students. (Potosnak, 1983, p. iii)

When people ask me about what I do, there are three things I commonly tell them about my field. First I say that "Human Factors Engineers are kind of like attorneys, representing humanity at the human-machine interface." Then I tell them that "we engineering psychologists have our fingers in the dike of rampant technology." Finally I boast: "My mentor is Alphonse Chapanis. I guess you could describe him as the 'Godfather' of Engineering Psychology!" This tells my listener that I have tremendous respect and affection for "The Professor" and that he, in turn has taught me to have respect and consideration for the "end users" of this world. If this paper eschews

105

sesquipedalian obfuscation, that is only because of professor Chapanis' patient and persistent efforts over the last four years to teach me to think and write clearly. For that contribution to my continuing edification, I maximally appreciate this opportunity to express my enduring gratitude...no No NO! What I mean to say is: "Thanks Professor, I am proud to have you as a role model!" (Kelly, 1983, p. iv)

Finally, I wish to express my sincere appreciation to my advisor, Dr. Alphonse Chapanis, whose guidance and infinite patience directed me through this doctoral program. To learn from him was a privilege, and to have been his last student will always be an honor. (Al-Awar, 1985, p. iii)

DISSERTATIONS

From 1968 on, I asked each student to provide me with a copy of his or her dissertation. I have had these bound for my personal library. Table 2 is a list of my students with the titles and dates of their dissertations. To make this list complete I should have added the names of many undergraduate students who worked with me, graduated, and then went on to graduate school in other universities, and the graduate students who worked with me for a while, but elected to do their Ph.D. dissertations with another faculty member. It was not my intent to slight them; I just could not recall all their names. A number of them are identified by name elsewhere in these memoirs and several are joint authors with me on articles listed in my bibliography.

Table 2. My students with the titles and dates
of their Ph.D. dissertations

Janan Arif Al-Awar: An Instrument for Evaluating the Use of Speech in Computer Communication, 1985.

Emanuel Averbach: The Apparent Brightness of Maxima and Minima in Luminance Gradients, 1956.

Mark Allen Brecht: Study of Meeting and Conference Behavior, 1979.

Charles Ray Brown: Difference Thresholds for Intermittent Photic Stimuli as a Function of the Rate of Flash, Number of Flashes, and Presentation Time, 1958.

William Randolph Ford: Natural-Language Processing by Computer – A New Approach, 1981.

Daniel Martin Forsyth: The Use of a Fourier Model in Describing the Fusion of Complex Visual Stimuli, 1959.

Bernard Adolph Gropper: Effects of Spatial-Temporal Ordering on Short-Term Memory for Arrays of Digits, 1968.

Rita May Halsey: A Comparison of Three Methods for Color Scaling, 1953.

Randall Melville Hanes: The Construction of Subjective Brightness Scales from Fractionation Data: A Validation, 1949.

Douglas Gordon Hoecker: Problem-Solving in Five Communication Modes as a Function of Verbal and Spatial Abilities, 1979.

Table 2. My students with the titles and dates
of their Ph.D. dissertations (continued)

John Falk Kelley: Natural Language and Computers: Six Empirical Steps for Writing an Easy-to-use Computer Application, 1983.

Michael John Kelly: Studies in Interactive Communication: Limited Vocabulary Natural Language Dialogue, 1975.

Gerald Peter Krueger: Conferencing and Teleconferencing in Three Communication Modes as a Function of the Number of Conferees, 1976.

Michael Leyzorek: Two-point Discrimination in Visual Space as a Function of the Temporal Interval Between the Stimuli, 1951.

Donald Aaron Mankin: The Influence of Perceptual Anchors and Visual Noise on the Vertical-Horizontal Illusion, 1968.

Paul Roller Michaelis: Cooperative Problem Solving by Like- and Mixed-sex Teams in a Teletypewriter Mode with Unlimited, Self-Limited, Introduced and Anonymous Conditions, 1978.

Robert Bruce Ochsman: The Effects of Ten Communication Modes on the Behavior of Teams During Cooperative Problem-Solving, 1973.

Charles Monroe Overbey: The Effects of Similar and Dissimilar Communication Channels and Two Interchange Conditions on Team Problem Solving, 1974.

Peter David Pagerey: Communication Control and Leadership in Telecommunications by Small Groups, 1980.

Table 2. My students with the titles and dates
of their Ph.D. dissertations (continued)

Robert Neil Parrish: Interactive Communication in Team Problem Solving as a Function of Two Educational Levels and Two Communication Modes, 1973.

William Thomas Pollock: The Visibility of a Target as a Function of its Speed of Movement, 1953.

Kathleen Marie Potosnak: Choice of Computer Interface Modes by Empirically Derived Categories of Users, 1983.

David Mark Promisel[9] : Visual Target Location as a Function of the Number and Kind of Competing Signals, 1960.

Joan M. Roemer: Learning Performance and Attitudes as a Function of the Reading Grade Level of a Computer-Presented Tutorial, 1981.

Tapas Kumar Sen: Visual Responses to Two Alternating Trains of High Frequency Intermittent Stimuli, 1963.

Alex Lewis Sweet: Temporal Discrimination by the Human Eye, 1949.

Carroll Vance Truss: Chromatic Flicker Fusion Frequency as a Function of Chromaticity Differences, 1955.

Gerald Dermot Weeks: Alternative Telecommunication modes for Conflictive and Cooperative Problem Solving, 1974.

Elizabeth Zoltan-Ford: Language Shaping and Modeling in Natural-Language Interactions with Computers, 1983.

---

[9] Recipient of D. Eng. degree

AN APPRECIATION

From force of habit I speak and write about *my* research as through it were something I alone had accomplished. Nothing could be further from the truth. I may have conceived of a study, but the hard work of carrying it out – testing subjects and analyzing the data – was done by my students. Even that is not entirely correct, because many good ideas for experiments came from them as well. Preparation of reports and manuscripts was always a joint process. Some times they prepared first drafts and other times I did, but the final draft was always done together. In fact, virtually everything I did from 1958 to 1985 was a cooperative effort. To a great extent I owe whatever success and prominence I may have achieved to them – my students. I could not have done what I did without them.

8. PRESIDENCIES
1959-1979

It has been my good fortune to have been elected president of three professional organizations. All three presented special problems and challenges but all were rewarding, albeit in different ways. In all three cases I was able to initiate changes in their methods of operation that, by the test of time, have proven to be successful.

THE SOCIETY OF ENGINEERING PSYCHOLOGISTS

Two things stand out in my memory about my 1959-1960 term as president of the Society of Engineering Psychologists (now the Division of Applied Experimental and Engineering Psychologists), Division 21 of the American Psychological Association. First, I started a tradition by giving a presidential address (Chapanis,

1961), the first ever delivered to the Society. My address was given on September 5, 1960, during the sixty-eighth Annual Convention of the American Psychological Association in Chicago, Illinois. At that time I had already been stationed in London for about three months and the second thing I recall is that I had difficulty persuading the Commanding officer, Captain J. K. Sloatman, Jr., and Scientific Director, Dr. I. Estermann, to give me time off and allow me to use MATS, the Military Air Transport Service, to return to the United States to give my address. They relented only when I said that I would take a commensurate reduction in salary and pay my own way on commercial air.

I had another more personal reason for returning to the United States. My son, age 12, had accompanied us on our trip to London in June 1960 and had spent the summer with us in London or traveling with us in Europe. I had to return him to his mother so that he could resume his schooling in September, and I was reluctant to let him return to the United States unaccompanied. At any rate, we both came back by MATS, I delivered my son to his mother, traveled on to Chicago, gave my speech, and returned to London shortly after.

My talk was well received and was subsequently reprinted three times (Marx, 1963; Fleishman, 1967). The third reprinting was as a separate (P-418) publication in the Bobbs-Merrill Reprint Series in the Social Sciences.

THE HUMAN FACTORS SOCIETY

My term as president of the Human Factors Society (now the Human Factors and Ergonomics Society) was much more eventful. I had been elected a member of the Executive Council in 1961 and president in 1963. At that time the Society was a fledgling organization with a total membership of a little over 1,100 persons and with annual revenues of roughly $15,000 and disbursements of $14,109 in 1963. There were no permanent officers, nor did the Society have a permanent home. Instead, files and archives, such as they were, were delivered from an outgoing elected officer to the newly elected one, wherever he or she might be. It was also loosely or casually governed, as the following incident will suggest.

> At the conclusion of the 1962 meeting held in New York City the outgoing president, Jack W. Dunlap, left before the annual business meeting. The newly-elected president, Paul M. Fitts, was not in attendance either. By default I was pressed into service to tell the assembled members what had transpired at the last Executive Council meeting. I don't know how I managed to survive that ordeal, but somehow I did.

Communicating with the Membership

When I assumed my office, I was determined to make some changes. I wanted the Society to be run in a more business-like and professional manner, and I wanted to do a better job of keeping the membership informed and involved. To the latter end, I initiated a series of letters called "FROM THE PRESIDENT...." which were printed monthly on the first page of the *Human Factors Society Bulletin*. The *Bulletin* at the time was a little 5" x 8" publication of from 4 to 10 pages, but it was the only way I had of communicating with the membership. My first Letter appeared in the October 1963 issue of the *Bulletin* (Chapanis, 1963b) and I was gratified that an editorial (p. 2) in the same issue commented "Many of us felt that this meeting was a milestone in the Society's development... The business meeting was well attended and lively and knew what it was doing. The Executive Council meetings were zealously and well conducted...perhaps...as some suggested...at the 1963 meeting, HFS came of age." I would like to believe that my influence was partly responsible.

I Appoint Marian G. Knowles Administrative Assistant

Until 1963 the Executive Secretary of the Society had been a volunteer, and, for the two years preceding my presidency, Charles W. Simon had been serving in that capacity faithfully and competently. When Simon announced his resignation to take effect at the end of the year, I took the bold step of hiring someone to serve that function. So in the December 1963 issue of the *Bulletin* (Chapanis, 1963c) I announced to the membership that I had appointed Marian G. Knowles Administrative Assistant in the Office of the Executive Secretary. It was a decision I've never regretted. For the next 26 years Marian dunned us for our dues, tracked us down when we moved and neglected to inform the Society, paid the Society's bills, recorded our actions, reminded us of things we had to do, became the repository of our traditions and history, and served as the font of knowledge not only for us but for persons outside of the Society who wanted to know who we were and what we did.

I Give the Society a Permanent Home

In the early years of the Society its official address was a corner of the Secretary's desk, and the Society's address moved with each change of officers. That way of doing business, I was convinced, had to change. The Society needed a permanent address. I remember the arguments that swirled around my proposal in our Executive Council. The cost would be prohibitive. We were a small society. Did we need to spend all that money? In the end, perhaps due to my persuasiveness, but more likely due to my obstinacy, we rented Suite C, 1124 Montana Avenue, in Santa Monica, for a three-year term, and in the March 1964 issue of the Bulletin (Chapanis, 1964a) I was able to announce to the membership that at last we had a home of our own. We've been there ever since.

I have also described both of these historical events in the Keynote address I gave at the 29th Annual Meeting of the Society (Chapanis, 1985).

I Start a Tradition

The last memorable thing I did for the Society was to initiate the tradition of an address by the outgoing president. Mine was the first, I believe, to call attention to the importance of words and wording in the signs, instructions, and manuals that are associated with the equipment we use (Chapanis, 1965b). It was gratifying to read this assessment of my talk in an article over 30 years later (Wogalter et al., 1997):

Chapanis' assertions of words and language in ergonomic design and research has had substantial impact. A casual glance at the professional human factors/ergonomics research journals or meeting proceedings over the past three decades leaves little doubt that the field has become intensely focused upon words and language as a critical tool

115

to benefit people's performance, satisfaction and safety... The direction of these investigations indicates that many of Chapanis' suggestions have taken root in ergonomic design and research. (p. 181)

My article also continues to be cited and used in other ways as is evidenced by a request I received in August 1997 from a professor Thomas Furness III, at the University of Washington, to reprint the entire article in 38 copies for a course he was teaching that fall.

THE INTERNATIONAL ERGONOMICS ASSOCIATION

The International Ergonomics Association (IEA) is primarily a confederation of federated or regional ergonomic and human factors societies, although there are provisions for other kinds of members, for example, individuals living in countries that have no federated society, honorary members, or corporations. The First Congress of the International Ergonomics Association was held in Stockholm, August 20-23, 1961, and was briefly reported in *Ergonomics* (1962, Volume 5, pp. 335-336). That issue of the journal also published the original articles of the IEA (pp. 333-334) and established *Ergonomics* as the official publication of the Association (p. 336), an arrangement that continues to the present.

According to a note in the 1968 (Volume 11, pp. 309-314) issue of *Ergonomics* I was elected a member of the Council in 1967. At that time, the following were the federated ergonomics societies (and their membership):

Ergonomics Research Society	475
Gesellschaft für Arbeitswissenschaft	162

Human Factors Society	1247
Japanese Ergonomics Research Society	208
Nederlandse Vereniging voor Ergonomie	119
Societa Italiana di Ergonomia	60
Société d'Ergonomie de Langue Française	121

Evidently the HFS joined the IEA sometime between 1965 and 1967.

A Symposium on Physiological and Psychological Criteria

At the Third International Congress on Ergonomics in Birmingham, England, September 10-15, 1967, and at the International Symposium on Ergonomics in Machine Design, held in Prague, Czechoslovakia, October 2-7, 1967, I was impressed by the physiological flavor of European research in ergonomics. I felt that in most cases the physiological criteria were inappropriate for design. So, in 1968 when I attended a meeting of the Council of the IEA during the annual meeting of the Ergonomics Research Society held at the University of Sussex, UK. I suggested that it would be interesting to take a serious look at the differences between American "human factors engineering" and European "ergonomics", particularly at their relative emphasis on, and use of, psychological and physiological methods and techniques (Singleton et al., 1971). This idea was accepted and led to an IEA sponsored symposium on "Physiological and Psychological Criteria for the Study, Design and Validation of Man-Machine Systems" (Chapanis, 1969a). The symposium was held at the International Centre of the Royal Tropical Institute, Amsterdam, The Netherlands, September 16-18, 1969. It was at that symposium that I presented my paper on "The Search for Relevance in Applied

117

Research" (Chapanis, 1971b).

IEA Dues

From the standpoint of the HFS, one of the principal obstacles to our joining the IEA had been the dues. These had been established at 2 Swiss francs for each active member of a federated society. Since the HFS had more members than all the other societies combined, the HFS was contributing more than half of IEA's income, but receiving no more benefits than the smaller societies received. I argued this point in IEA Council meetings (*Ergonomics*, 1974, Volume 17, pp. 569-575) and finally managed to get the assessment fees to be made commensurate with the size of a society (*Ergonomics*, 1977, Volume 20, pp. 432-433).

A Tribute to my Service

Since Council members were allowed to serve only two consecutive three-year terms, my tenure ended in 1973. The Minutes of the 1973 meeting of the Council (*Ergonomics*, 1974, op. cit.) had this to say about my tenure:

> Prof. Chapanis had served the I.E.A. for many years now as the only American Member of Council, and also as an officially appointed representative of the Human Factors Society in connection with the preparation of the 1976 Congress. In spite of the distance between Europe and U.S.A. and the fact that most Council meetings were held in Europe, he never failed to participate. He had contributed much to the development of the I.E.A. and had

taken the initiative for the last two I.E.A. symposia in the recent past. (p. 575)

Some Organizational Changes

Some intimation of the changes I would try to bring about is contained in those same minutes: "Chapanis added that the existing I.E.A. structure was weak and this could be demonstrated in many ways. In his opinion a special committee should be elected by Council. For instance there were no rules to change the Articles" (p. 574). One change I instituted was secret balloting to replace the casual "Ol' boy" way of electing officers and Council members. This seemed to run counter to the European way of doing things, but the change was finally made.

Another change I initiated was the deliberate solicitation of contributions from corporations as a way of increasing income to IEA. Again, this ran into surprising opposition with arguments that it wouldn't work in Europe, that it wasn't a dignified thing for the organization to do, that it wasn't even worth trying. Nonetheless, I persisted and with the help of my treasurer, Herman Scholz, drafted a letter in German which I sent to a few selected industries. I felt vindicated when we received our first contribution from IBM Europe in 1977, I believe. These days, the IEA has even gone so far as to solicit bequests to the IEA.

A Permanent Address

As was the case with the HFS when I assumed the presidency of that society, the IEA had no permanent address. In fact, in the late 1970s the HFS was the only federated society that had a permanent address. For that reason I recommended to the Council that the IEA use the HFS office as a stable post office. My suggestion met with almost immediate and strong opposition because most European members of the Council feared that the IEA would be identified with HFS or that our Society would overwhelm the IEA and compromise its international flavor. I argued that my intent was merely to provide a post office address from which correspondence could be forwarded to the appropriate IEA officer or federated society. In the end, good sense prevailed. Mail addressed to the HFS/IEA could be sure of reaching its intended recipient and that is still the case.

I Create a Theme and Logo

In our planning sessions for the 1979 Congress we were searching for a theme. One evening, it came to me: "Old World, New World, One World"! Along with that I drew up a logo (Figure 6). It pleased me that my theme was adopted by the planning committee and that my logo was accepted as the official IEA logo (Ergonomics, 1977, Volume 20, p. 434).

Figure 6. The Logo of the International Ergonomics Association.

THEN... AND NOW

The Society of Engineering Psychologists (SEP, now the Division of Applied Experimental and Engineering Psychologists) has had a strange and precarious existence. In its early days, up until about 1965, it sponsored sessions at the annual meetings of the American Psychological Association (APA) that were interesting and well attended. Then as the membership of the APA grew and the number of divisions exploded, interest in Division 21 and attendance at its meetings declined, primarily, I believe, because the division was being overwhelmed by sheer numbers and squeezed out of the time slots for its own meetings. The last APA meeting I attended was in 1979 and I recall being lost in the throngs of people with whom I had no community of interests. No wonder. At that time Division 21 was one of 37 APA divisions and the number of its members was just 1% of the total APA membership. By 1998 Division 21 had become even more forlorn. It was now only one of 49 divisions with a membership less than 0.5% of the total number of APA members.

I also find its new title, the Division of Applied Experimental and Engineering Psychologists, strange because applied experimental psychology, even though we use it in our first textbook (Chapanis et al, 1949), does not define a clearly circumscribed area of interest. Nonetheless, vigorous efforts are being made to rejuvenate the Division. Although I find that the division no longer serves my interests, I am proud to have served as its president and I wish it well.

The Human Factors and Ergonomics Society (HFES) and International Ergonomics Association (IEA), on the other hand, have thrived and grown into well-known and highly respected organizations. I would like to believe that the innovations I instituted in both organizations during my terms as president have contributed to their growth.

9. A YEAR IN EUROPE
1960-1961

From June 1960 to September 1961 I was a Liaison Scientist in the Office of Naval Research Branch Office in the US Embassy, London. Although the office in which I worked was officially part of the Embassy, it was located physically in a building on North Audley Street, a few blocks from the Embassy on Grosvenor Square.

MY DUTIES

It was a great year. My duties were to (1) visit European institutions and write reports about what I had seen for distribution to the United States, (2) advise American scientists who came to Europe for official visits, meetings or conferences, (3) advise European scientists who wanted to visit the United States, (4) entertain foreign visitors to the UK, and (5) distribute certain reports from the US to selected foreign scientists.

REPORTS

We prepared two kinds of reports. *Technical Reports* were long and detailed about some one topic or visit and were prepared by a single author. Shorter contributions, ranging in size from a single paragraph to three or four pages, were combined with contributions from other Liaison Scientists into a publication called *European Scientific Notes*. All reports carried the following notice on their covers: "This document is issued for the information of U.S. Government scientific personnel and contractors. It is not part of the scientific literature and must not be cited, abstracted, reprinted, or given further distribution." For that reason none of those reports is cited in my bibliography.

Altogether I authored seven *Technical Reports* and contributed to 19 issues of *European Scientific Notes*. The most memorable and interesting one in my opinion is a 21-page Technical Report (ONRL-36-61) titled "Observations on some psychological activities in Israel." As an official representative of the U.S. Government I was received royally and had the opportunity to visit and learn in detail about Kibbutzim, the psychological profession in Israel, the Armed Services (Army, Navy and Air Force), the Israel Institute of Applied Social Research, and the Hebrew University. The truly lasting impressions that I carried away from that trip, however, were the incredible sights and sites to which I had been treated during that visit.

A Special Evaluation Report

On at least one occasion I was requested, through proper channels of course, to evaluate certain special European programs. The trip that I made for that purpose did not come out of ONR's budget, but was financed by the agency that requested the evaluation. The report I wrote on that assignment was not part of the regular ONR series, but was sent back directly to the agency that requested it.

PEREGRINATIONS

I carried a special passport which smoothed my passage through immigration and customs whenever I traveled, except for Italy where an officious customs inspector in the Rome airport once refused to respect my special passport and insisted that I pay duty on some slides I was carrying to illustrate a lecture. Although he finally relented after consulting with a superior, that incident colored my attitudes toward the country which even now I find difficult to overcome. I traveled widely: Belgium, Denmark, France, Germany, Greece, Holland, Israel, Italy, Sweden, Switzerland, Turkey, and the UK, of course. Cram courses in French at the Embassy in addition to collegiate courses left me fairly proficient in that language, an asset that proved itself many times and sometimes in unexpected places, Turkey, for example. German, on the other hand, was more useful in Greece and Israel. All in all, it was one of the most exciting years in my entire professional career.

MY FORAY INTO SOCIAL SCIENCE

While I was working at the Embassy, my wife was employed at the Tavistock Institute in London. One of her projects was to review and critique literature on cognitive dissonance. When we would sometimes meet for lunch on the grass in Hyde Park she would bring an article from the literature for me to comment on. I would often find methodological flaws, unwarranted interpretations of the data, and faulty reasoning in these articles. There were so many of these that she finally prevailed upon me to join with her in writing an article titled "Cognitive Dissonance: Five Years Later" (Chapanis & Chapanis, 1964). The article, but with a different title, "A critical evaluation of the theory of cognitive dissonance," was reprinted in Southwell and Merbaum (1964). This was my one reluctant foray into social science, and I have always thought of it as a rather incidental piece. Yet that article, I understand, created quite a furor when it appeared, and I find that it is even cited in some current literature. I am amused that in some circles when psychologists hear my name, they associate me with that article and not with human factors.

SO?

So, what good did it do human factors? Despite everything I learned about Europe, its culture, and European ergonomics during that year, I cannot in all honesty claim that my tour of duty did much to advance the human factors profession. My reports went to a select list of recipients and I never knew who was on that list. Moreover, I never received any feedback from anyone about any of the Technical Reports or European Scientific Notes I wrote. For a few years after my return to the United States I received reports written by my successors, but never found them interesting enough to preserve.

Perhaps the greatest benefits to have come out of that experience were that I met and got to know many professional people I would meet many times afterward and that many Europeans got to know me and my work better. These mutual acquaintanceships served me well when, for example, I became active in the affairs of the International Ergonomics Association (Chapter 8) and organized special international meetings (Chapter 12).

10. THROUGH
THE IRON CURTAIN
1967

In the summer of 1967 I had a harrowing experience which I am able to recount accurately and in some detail because I wrote detailed notes about the events of that summer and those notes somehow escaped many trips to an incinerator. Although it was neither the first nor the last time I penetrated the Iron Curtain, it was emphatically the one that left me with the most chilling memories.

PRELUDE

Late in 1966 I knew that I would be attending the Third International Congress on Ergonomics in Birmingham, England, in September 1967. In fact, I had been working on the theme address I was planning to give at those meetings (Chapanis, 1967). Then early in 1967 I received an invitation from the International Labour Office in Geneva to present a paper at a symposium to be held in Prague, Czechoslovakia, from the 2nd to the 7th of October, 1967.

The invitation sounded intriguing but it was to be held in a country behind the Iron Curtain and I held a SECRET security clearance which conceivably placed me at some risk. Consultations with Dr. Gilbert C. Tolhurst, Head of the Physiological Branch of the Office of Naval Research, and with Mr. Alfred Ashton, head of government security at Johns Hopkins, resulted in permission to attend the symposium after appropriate briefings from Mr. Ashton.

SOME PROFESSIONAL VISITS

After attending the meetings of the Congress in Birmingham (September 10-15, 1967), I flew to Brussels, Belgium, and picked up a car (a sports Volvo 1700) which I had ordered from the United States before leaving on my trip. Since the Prague symposium was three weeks away I visited some European laboratories in the interim. My first visit was to the National Defense Research Organization TNO, Institute for perception RVO TNO, Soesterberg, The Netherlands, where I saw a number of very good studies and where I gave a talk (subject matter forgotten) about some of my research.

A Visit with Jacques Leplat

My second visit was to the Laboratoire de Psychologie du Travail, of the École Pratique des Hautes Études, 41 Rue Gay-Lussac, Paris, where I was entertained by Jacques Leplat, an old friend. There I once again saw some innovative research. My notes say that at the conclusion of my stay, Leplat had invited a group of about 30 psychologists and human factors engineers from Paris and its suburbs for a seminar. I talked – in French – about some of my research. Since my French was not eloquent, we had a hilarious time. My notes also say that "The seminar was followed by a magnificent French dinner, served with typical Gallic good spirits and conviviality."

For some reason my car was delivered with German license plates. Since I planned to go into a country that was hostile to Germany, I took the precaution of going to an AAA office in Paris to get USA decals which I affixed to the front and rear of my automobile.

And then with Etienne Grandjean

My third visit was to Professor Dr. med. Etienne Grandjean, Director of the Institut für Hygiene and Arbeitsphysiologie der Eidgenoessischen Technischen Hochschule, Classiusstrasse 25, Zürich, Switzerland. I had met Grandjean earlier in Ljubljana, Yugoslavia. My notes say that his "institute conveys the impression of great vitality, orderliness and industry. In comparison with my own office, Grandjean's is huge. It is well organized, tidy, and tastefully arranged. His laboratories seem large and well supplied. Much of the laboratory apparatus appears to have been made with the precision of a Swiss watch. But entirely aside from such material considerations is the research output of the institute... This is clearly a productive team. Equally impressive is Grandjean himself. He writes and converses easily in French, German and English and speaks still another language,

Italian!"

THE JUMPING OFF PLACE

After leaving Switzerland, I traveled through western Austria and stopped overnight on Saturday, September 30, in Linz, a small Austrian city about 50 kilometers from the Czech border. Sunday morning, October 1, was a beautiful day. Church bells were pealing and the citizens were strolling about in colorful costumes. After an early Frühstück, I packed up and started on the road. The road passed through an idyllic countryside and after I passed the small village of Freistadt, the last before the Czech border, I had an eerie feeling. Then it struck me: I was the only vehicle on the road! There was no other vehicle or person to be seen.

THE IRON CURTAIN

Passing through the Austrian border station was almost perfunctory. But shortly thereafter I came to a soldier standing in the middle of the road with a machine pistol cradled in his arms. He raised his hand commanding me to stop. Off to one side among some trees was a tank with its cannon trained on the spot were I had stopped. The soldier slowly and deliberately walked around to the car, peered underneath, and said something which I took to be a request for my passport. He studied it carefully, comparing the photograph with my face. Apparently satisfied he handed my passport back and examined the interior of the car. Finally, after a tense 15 minutes or so, he motioned me to proceed.

Several hundred yards later I came to a border barrier. Soldiers lifted a barrier and opened a gate allowing me to drive into an area completely enclosed with fencing. Another tank guarded the area. Here the real examination began.

Automobiles manufactured in Europe at the time had standards for headlights that differed from American standards, and when I accepted my car in Belgium I was given a cardboard carton with two headlights that met American standards and told to have them installed after I returned home. That carton in the trunk of my car caused the border guards no end of concern. In halting but understandable English I was asked: What are these? Why am I carrying them in my car? What am I going to do with them?

Finally satisfied, the interrogation turned to me and my visit. Where am I going in Czechoslovakia? Why? What am I going to do? How long will I be in the country? How much money am I carrying? Show me your money. Slides that I was carrying to illustrate the talk I gave in England and the talk I planned to give in Czechoslovakia had to be examined one by one while I tried to explain what they were about. My passport meanwhile had been handed in to someone in an adjacent building and I wondered if I would ever see it again. After about a half hour of this searching examination, I had an idea: Show them the official letter of invitation from the International Labour Office (ILO). That, too, disappeared in the adjoining building but appeared about ten minutes later with my passport. The second wire fence was moved aside and I was allowed to proceed. The entire episode had been so tense that at one point I had almost decided to turn around and return.

THE SYMPOSIUM

That hurdle passed, I arrived in Prague without any further

incident. The symposium was held in the International Hotel and had been jointly organized by the ILO and the Czechoslovak Medical Society J. E. Purkyne. The hotel was close to the Russian Embassy where armed soldiers patrolled the streets surrounding it. I gave a paper on control-display movement relationships for right- and left-handed operators (Chapanis & Gropper, 1968). I also chaired a session of papers and was asked to make a little speech of thanks to the organizers of the symposium at the concluding session. I believe I was the only participant from North America (Mexico, Canada and the US) at these meetings.

Figure 7. Conferees at the Symposium on Ergonomics in Machine Design, Hotel International, Prague, Czechoslovakia, 6 October 1967. I am at the center of the photograph.

A Strategy Session

At the Birmingham meetings three weeks earlier, Dr. H. P. Ruffel Smith had expressed annoyance that this symposium should have been held so soon after the IEA meetings. To follow upon this matter, Dr. Etienne Grandjean, Secretary of the IEA, arranged a luncheon at which he, I, Drs. L Parmeggiani and G. Lambert of the ILO, Dr. Martti J. Karvonen of WHO, and Professors Alain Wisner and Erling Asmussen, also from the IEA, discussed the situation. The ILO wanted to have a symposium like this one every year, whereas the IEA has a Congress only every third year. To delimit the spheres of interest of these two organizations we decided that future meetings of the IEA should be primarily technical and scientific whereas future meetings of the ILO would be more practical and mostly concerned with the applications of ergonomics and with the education of ergonomists.

Some Shady Dealings

Every day little incidents constantly reminded me that I was in a Communist country. Two, however, stood out in my memory.

One evening a friend and I decided to escape the pedestrian fare served at the hotel and its environs by having dinner at what was reputedly the best, and most expensive, restaurant in Prague. Although somewhat faded, it did indeed retain vestiges of prewar elegance. There were not many patrons in the restaurant, but both the meal and wine were excellent and the service, for a Communist country, attentive. When it came time for our check, the waiter said in broken but understandable English and in a low voice, "American? Dollars?" "How many, " I asked. "Ten," he replied. I countered with "Five?" to which he nodded assent. I pulled out my wallet and was about to hand him a five dollar bill when he whispered urgently, "No, no, under table. Police," indicating by a small gesture a man seated alone in a corner behind him. So I placed the five dollar bill under my plate and we left. He followed us to the door bowing and murmuring his thanks in Czech.

The second incident was quite different. One morning a rather attractive lady accosted me and introduced herself as a reporter for a local newspaper. Saying that she wanted to interview me, she suggested moving to the hotel lounge for a drink. After getting settled, she made some polite inquiries about me, my profession, and life in America. Then she said that she had close relatives in America with whom she wanted to communicate, but could not because of government restrictions. Would I please take a letter with me and mail it to her relative after I had left the country?

This was one of the situations about which I had been expressly warned in my pre-trip briefing. If I accepted such a letter my belongings and I would be searched on leaving the country, the letter would be discovered, and upon being opened would be found to contain secret information, thereby providing the rationale for my arrest. So I politely refused and kept refusing despite her repeated and urgent pleas. I will never know, of course, whether she was a plant, but the whole situation struck me as being false and contrived.

RETURN TO FREEDOM

On leaving I chose the shortest route to West Germany. This took me through Cheb in Czechoslovakia and Marktredwitz in Germany. At the Czech border station I passed through barriers like those I had encountered on my entrance. My baggage, car, and I were indeed searched, but since I had been certain to take nothing illegal I was allowed to pass.

When I came to the German boarder post, marked by a building and a flag, there was no one outside. I stopped, went into the building and found two German border guards watching a soccer game on television. I inquired of one if he wanted to see my passport, but with a wave of his hand he indicated that I should just drive on. It was clear that he couldn't be bothered. And it was then I knew that I had returned to the free world. Well, not entirely, because after I had made arrangements for my car to be shipped home, I flew into West Berlin and there encountered the wall and "Checkpoint Charlie" through which one went from West to East Berlin. Ominous and frightening. It was good to get back to the USA.

11. CONSULTING
1974-1996

I mention or allude to various consulting activities in other chapters of this book. In this one, however, I describe activities that do not seem to fit elsewhere and make some general observations about the consulting business.

SOME PEDESTRIAN JOBS

Although some of the best human factors work I ever did was as a consultant, some of my jobs were humdrum and not worth much attention. For example, years ago when hand-held devices were just becoming common I made some simple simulation tests to arrive at an appropriate weight for hand-held inventory computers – the kind you see grocery clerks using when they go through the store checking the shelves. I can no longer remember the company for which I did this work nor can I remember the figure I arrived at. But that's OK. It's not really worth remembering.

As another example, in 1977 I prepared an 18-page report for Planning and Human Systems, Incorporated, a small organization in Washington D.C., on "Human Factors in the Use of Alphanumeric Codes." My report provided supporting material which that organization needed for its contract to develop specifications for VINs – Vehicle Identification Numbers. I kept a copy of my report because at one time I thought I might amplify it into a publishable article. But in reading it today I see that it really is another ho-hum job – neither exciting nor intellectually taxing.

AND SOME DISAGREEABLE ONES

I also had some unpleasant experiences – introductions to the real and sometimes unethical and unscrupulous practices of business and commerce. The ones that irked me the most were instances in which firms or other organizations invited me to be a consultant and requested a vita which they incorporated in their contract proposals. Then later, sometimes years later, I would find that they had been awarded the contract and made no effort to use my services. They obviously used my name and vita to enhance their contract proposal but never had any intention of using me.

This practice backfired on one occasion I know about. After one organization submitted a final report, the contract monitor, obviously dissatisfied with it, asked the organization to submit a letter from me confirming that I had followed good human factors practices in my work on the project. Needless to say, I took great delight in replying "X#§@% you!" I hope that organization was never paid.

Human Factors Versus the Market Place

I recall clearly another different kind of incident because it was so dramatic and because I recounted it to each new class of my students. I was invited to be a consultant to the Chrysler Corporation a few years after Lynn Townsend had been made Chairman of the Board in 1967. Shortly after I met with members of the design department, it became clear that they were not interested in my advice. They merely wanted me to give my approval to their designs – to give them a human factors blessing, as it were. I spent a great deal of time trying to convey to them that human factors was not just common sense and that I could not say whether a particular product was, or was not, well human-engineered merely by looking at it. I'm not sure whether I was successful.

At one point, I was taken to Townsend's private show room where the newest models of Chrysler's products were on display. I was looking at a sporty model that had a steering column with a sharply pointed tip extending an inch or two beyond the steering wheel. Townsend asked me what I thought about it. My exact, or very nearly exact , words were, "Mr. Townsend, do you know what you've designed here? You've designed a spear aimed at the driver's heart." I also remember distinctly his cynical reply, "Doc, it'll sell."

Experiences such as these were fairly common in the early days of human factors, but, I'm happy to say, I encountered fewer and fewer of them as the discipline became better known.

SOME DIFFERENCES BETWEEN
THE UNIVERSITY AND INDUSTRY

Very early in my consulting I was impressed by some very basic differences between the philosophy underlying research in a university and work done as a consultant.

Ownership of Intellectual Property

In a university, one tries to be scrupulously correct in acknowledging authorship and the contributions of anyone who assisted in or contributed to a study or an article. In consulting with industry, I found that I had no rights of ownership in anything I did. Industrial philosophy is that a consultant is hired to do a job and whatever the consultant produces belongs to the industry and not to the consultant. That was the case with the patent I was granted with Rust and Feagin when I was consulting with the Humble Oil and Refining Company (Chapter 3, p. 42-43). A number of other instances are cited later in this chapter and in Chapter 12.

Further, in a university setting the goal is to publish and make your results known to the world, because that is where your rewards lie. Industry, by contrast, feels that it pays for research to gain a competitive advantage and so has no incentive to make research findings public.

As a result of these policies my name is not associated with any of the products that resulted from my consulting activities. In some cases I was not even allowed to see the final report or the final version of the product I worked on, and in most cases I have nothing to document the work that I did. For the most part I can't say that I am resentful. After all, I was paid, and in most cases paid well, I had fun doing the work, and I had feelings of accomplishment even if I had nothing in writing to show for it.

Generalizations Versus Project-Specific Specifications

Perhaps the greatest difference between the two arenas is that in a university setting one does research to arrive at generalizations, general principles, and, if possible, to fit results in a theoretical framework. In fact, the greatest rewards, the highest recognition, goes to the researcher who can propose or concoct a theory. Even if the theory is later proven to be wrong, the person who formulated it will gain a certain amount of prominence because his or her name will be cited by everyone who disputes the theory or attempts to prove it wrong.

Industry, I found, has absolutely no interest in theories and generalizations. When you are hired as a consultant it is to do a job, to find out how to do something or do something better. Repeatedly I heard statements like, "Don't give me a theory or generalizations. Tell me exactly what I have to do." That is the reason why general rules and guidelines, which human factors loves to produce in great quantities, are often completely disregarded by engineers and designers.

The same applies to many human factors specifications. The following is one of many mandatory specifications I could have cited from what is unquestionably the most important and widely used human factors standard (MIL-STD-1472E, 31 October, 1996): "The [computer] interface shall optimize compatibility with personnel and shall minimize conditions which can degrade human performance or contribute to human error" (p. 161). Although one cannot argue with the intent of that specification, it does not tell the engineer or designer exactly how to "optimize compatibility with personnel" or how to "minimize conditions that can degrade human performance". In other words, it leaves it up to the engineer or designer to do the job that human factors should do. Moreover, it does not make it possible for an inspector, who most probably is not a human factors professional, to decide whether the interface has or has not optimized compatibility with personnel or whether the interface does or does not minimize conditions that can degrade human performance.

What the engineer or designer needs are project-specific specifications, not general rules. This is something many human factors personnel do not clearly understand.

PERSONNEL RESEARCH
FOR THE MERCHANT MARINE

In 1973 Operations Research, Inc., an organization then located in Silver Spring, Maryland, received a contract (Contract DOT-CG-31571-A) from the U.S. Coast Guard to develop and demonstrate a methodology for evaluating the effects of various actions the Coast Guard might take to reduce the risk of oil or hazardous cargo spills resulting from ship and barge collisions. Recognizing the importance of the human factor in these accidents, ORI engaged me to prepare a program of research on personnel qualifications, training, and licensing for the improvement of merchant marine safety. I, in turn, engaged Bernard A. Gropper, one of my former students (see Table 2, p. 107) to assist me.

As part of our preparation Gropper and I spent some time observing ship maneuvers and docking operations in the Port of Baltimore and interviewing marine pilots. I also visited the ship simulator facility at Kings Point, New York. Gropper and I then prepared what I now find was a comprehensive report covering such topics as the system concept, a comparison of aviation and marine personnel practices, job and task analysis, structuring an educational curriculum, course content, international considerations, teaching methods, training and research simulators, examinations, performance evaluation, licensing, program continuation, skill maintenance and enhancement, and periodic reevaluation of the personnel subsystem (ORI, 4 March 1974). I don't know whether our recommendations were implemented.

I INCORPORATE MYSELF AS
ALPHONSE CHAPANIS, PH.D. PA

In the 1970s IBM was promoting something it called *word processing* but was finding a great deal of resistance to its acceptance. In 1973 Dr. Bret Charipper of the Office Products Division, IBM, approached me with the suggestion that I undertake a research study of word processing. The negotiations with IBM were a little unusual. For some reason, their contracts personnel objected to a contract with the University. They were, however, willing to give the money to me. I consulted with my lawyer who pointed out great difficulties in my accepting these funds as an individual. If I were a corporation, however, my corporation could accept and use the funds without any risk to me as an individual. So in 1974 I became a corporation, with an accountant, separate bank account, corporate seal, and all the trappings that are associated with a corporation. (The PA, incidentally, is an abbreviation for "Professional Association.")

As it turned out that was a wise move. All my consulting income from that time on was funneled into my corporation and not to me as an individual. It greatly simplified my income tax returns because my University income was personal and separate from my consulting income. Incorporation also had a number of unanticipated benefits that are not worth detailing here.

STUDIES OF WORK ACTIVITIES IN OFFICES

Once incorporated, IBM's Office Products Division gave me what for those times was a generous grant ($135,000 as I recall) to (1) develop instruments for measuring work loads, work flows, and task complexities in office environments, (2) measure work loads and work flows in offices with and without word processing, and (3) develop a system model for evaluating the true costs of office systems.

This resulted in a 154-page report (Weeks at al., 1975). It is a good report. Among other things, we developed a system of categories for making activity analyses of office work and with that system studied two large offices, one with word processing, and the other without. One of our findings was what we called job displacement, a displacement of lower level activities – filing and telephoning, for example – onto higher level personnel in offices with word processing. Such displacement is commonplace these days. Typing skill is almost a job requirement for engineers, managers, and other highly-paid professionals who now use keyboards to communicate, make their own appointments and travel arrangements, and do other tasks that were once performed by lesser-paid secretaries.

When we completed our study, as a matter of courtesy we reported our findings to the office managers. In both cases, the managers said something like, "This is the first time I have been able to understand what people are doing in my office."

We also developed a system for rating office job complexity and provided benchmarks for each level. This report was accompanied by another (Chapanis, 1976a) that identified the elements that go into any realistic evaluation of the costs of any office system and showed how these costs should be combined to arrive at a final cost figure for the system. Out of all this work we were allowed to publish only one minor methodological paper in the open literature (Hartley et al., 1977). All that good work and so little to document what we did.

A STUDY OF COMMUNICATION CHANNELS

About 1977 IBM had teamed up with Satellite Business Systems (SBS) to use satellites for company communications. I was approached by an IBM engineer, Rick Melzig, to design a study to (1) compare communications relayed through a geostationary satellite with those relayed through regular telephone long lines and (2) compare the intelligibility of digitized speech with regular telephonic speech. Two test sites were to be used: one in IBM's corporate headquarters in New York, the other in San Jose, California. Test subjects were to be IBM managers.

The Design of the Study

For this study I wanted to have test materials that would be sensitive enough to detect small differences in speech intelligibility and yet be viewed as realistic by IBM employees. What I conceived of was lists of names, addresses, and social security numbers (actually I gave Joan Roemer, one of my graduate students, the work of compiling them). These were to be read by one talker and recorded by a listener at the other test site. In addition, I prepared a complete test protocol with instructions that were to be read to the subjects, the sequences in which communication channels and subjects were to be tested, and how the results were to be scored. Subjects, of course, were never to be informed about what kind of a channel of communication they were using.

My test protocol was apparently very successful because the tests were carried out without a hitch. Some months later, Melzig brought the data back to me for analysis. The analysis showed, as I recall, that digitized speech via a satellite was more intelligible than speech transmitted over ordinary telephone long lines. This was a highly satisfying job, satisfying because it was done to answer a practical design question, it used real equipment, it used subjects who were truly representative of the using population, and it resulted in a usable answer that satisfied my client. Moreover, I felt I had planned and handled the study capably. My name, however, was never associated with this project and I was never even allowed to have a copy of Melzig's final report. I have always regretted that I was never allowed to publish this study. It would have made a neat article in *Human Factors* or *Applied Ergonomics*.

I AM CONSULTED FOR THE WRONG PROBLEM

About the time that I completed my work with IBM/SBS, I was asked to consult about a problem that the director in our School of Public Health was having with a system called POPINFORM. This was a computerized, on-line information system that provided citations and abstractions of journal articles, book chapters, published and unpublished reports, government and United Nations documents, and other population and family planning data. It was a worldwide system with remote terminals in many foreign countries, designed to be used by public health personnel, physicians, and government officials. The system was supported by AID, the US Agency for International Development. Although a computerized system, it did not have a visual display terminal. Instead, queries were input on a conventional keyboard and results were printed out on paper.

A Motivational Problem?

When Gerald D. Weeks and I were introduced to the system, the director said her problem was a motivational one. What could be done to motivate public health officials in places like new Delhi, Sri Lanka, and Manila to use the system? Perhaps it was, we suggested, that the users didn't speak or understand English. We were assured that was not the problem, because all the users had received their medical training in the US. I was about to leave because I told the director that we were not experts in motivation. Before doing that, however, I asked if I might use the system, and after some gentle insistence was allowed to do so.

The Culprit: User Unfriendliness

Almost immediately, I was lost in a maze of incomprehensible abbreviations, esoteric computer jargon, and enigmatic responses. After a dozen or so interchanges I gave up. My analysis of what had happened finally convinced the director that the problem was almost certainly not a motivational one, but rather a user unfriendly computer. If I, an American with some computer experience, couldn't make any sense of the interchange, how could she expect acceptable performance from persons with perhaps a less complete understanding of ordinary English? My examination of the user manual showed that it was equally poor and unhelpful. As is so often the case, it had never occurred to the director that the computer was difficult to use.

This was such an interesting incident that I preserved several pages of interactions and reproduced them faithfully in an article I wrote (Chapanis, 1982a, pp. 126 & 127) and in an invited address I gave at a Human Factors Symposium sponsored by the ITT Europe Engineering Support Centre and held in the Excelsior Hotel, Heathrow, London, on May 18 and 19, 1982 (Chapanis, 1982c p. 37).

ROLM CORPORATION'S PHONEMAIL™ SYSTEM

At about the same time as I had the experience described above I served as a consultant on Rolm Corporation's Phonemail™ system. Specifically, I helped to develop the decision tree, the dialogs and prompts, labels for keys, and the wording used in the system. Early in the project I made one very important change in the way alternatives were presented. The engineers originally had prompts that said in effect, "Press key W if you want to do X, press key Y if you want to do Z,... I convinced them that the prompts should reverse the order and say "If you want to do X, press key W; if you want to do Z, press key Y;..." Impromptu tests showed that my ordering was easier to use. When I returned to Baltimore, Rolm connected me to their system so that I could use it for whatever additional tests I wanted to make.

After the product was put on the market, I was gratified to read in the December 1982 issue of *Datamation* the following:

I have concluded that the Rolm product is likely to be superior in functionality to any other voice mail system currently available, says Bruce Hasenyager, a vice president of Kidder, Peabody & Co., who has followed the voice mail market for some time. They've done a good job of designing the product from a human factors standpoint. (Tyler, 1982, p. 54)

I was even more gratified when IBM acquired the Rolm Corporation in 1983 and adopted Phonemail™ instead of IBM's own voice mail system. I discussed this work briefly in Chapanis (1991c) and I illustrated Phonemail™'s keyset and decision tree in lectures I gave in the US and Europe in the World Conference in 1984 [about which more later].

A STUDY OF CHESAPEAKE BAY PILOTING OPERATIONS

Ships leaving the port of Baltimore are guided by a pilot down the length of Chesapeake Bay (a little over a hundred miles) to Cape Henry where they enter the open waters of the Atlantic Ocean. At Cape Henry pilots are transferred to a pilot ship, a kind of dormitory ship with a fully equipped kitchen staffed around the clock (because ships arrive and depart at all hours of the day and night). After at least six hours of mandated rest, pilots would pick up another ship entering Chesapeake Bay and guide it back to the port of Baltimore. It is a demanding task because the pilot must be thoroughly familiar with the features of the Bay throughout its entirety and must be able to recognize and identify channels, waterways, bridges, obstructions, buoys, and other vessels by lights and other features under all conditions of illumination and in all kinds of weather.

In addition, every vessel is unique with its own steering and handling characteristics. Moreover the pilot must be prepared to contend with inadequately maintained or malfunctioning equipment, poorly trained crews, and imperfect communication with persons of diverse nationalities. Finally, there is the almost constant threat of a collision or grounding.

The Problem: Pilot Fatigue

In 1979 the Association of Maryland Pilots approached me to study the problem of fatigue among pilots. In preparation for the study I interviewed pilots, studied the office operations in Baltimore, made a trip from Baltimore to Cape Henry on an outbound vessel, stayed overnight on the dormitory ship, and took the next inbound vessel back to Baltimore.

Among other things, I devised a Trip Log to collect 12 pieces of data for each trip: the time the pilot received his first call, the ship's scheduled departure time, the time the pilot arrived at the pilot office, the time the pilot boarded the ship, the time the ship actually departed, and so on. Some of this information was available from logs kept in the pilot office; some was recorded by the pilot. I also prepared an Individual Work Schedule form on which each pilot recorded for each day and for each outbound trip (1) the time of first call, (2) the time of arrival and the pilot office, and (3) the time of disembarkation. The same form called for (1) the time the pilot boarded a return ship, (2) the time the pilot arrived at the pilot office in Baltimore, and (3) the time the pilot returned home. In one month, August 1980, I had data for 396 trips. From computer printouts of time usage reports I also had data on 5518 bay transits between October 1, 1979 and September 30, 1980.

For this study I hired Mr. John R. Erickson to oversee the data collection, Mr. Jeffrey P. Schwartz, a graduate student, to do library research, and my wife, Natalia P. Chapanis, to do the statistical analyses of the data.

Findings

My findings were several (Chapanis, 1980c). Most important was that bay trips were excessively long. The median trip took 10 hours, about 10 percent of trips took longer than 13.5 hours, and some trips took as long as 49 hours! In addition, I found that work loads were distributed very unevenly among the various pilots and that the daily and weekly work schedules of individual pilots were highly irregular.

Recommendations

My most important recommendation was that a mid-bay station be established so that the Baltimore to Cape Henry run could be split into two segments with a different pilot taking each segment. This recommendation was eventually implemented. Another recommendation was that steps be taken to distribute work loads more equitably and to achieve more regularity in the work schedules of individual pilots. Finally, I recommended that the State of Maryland Board of Examiners of Maryland Pilots should promulgate regulations governing the hours of work, work schedules, and rest and recovery times for piloting operations that are consistent with Federal regulations for operators of air and highway transport vehicles. This also was finally done.

An Evaluation

This study was not an experiment; it merely involved the collection, tabulation and analysis of observations. Nonetheless I am proud of it and feel that I did a good job in attacking a serious practical problem. Because the data involved matters of government regulation, a few years later I was called upon to testify about it before the Maryland Board of Examiners of Maryland Pilots and, of course, I was not allowed to publish an article about it, although I did use one figure from it (Figure 530, p. 201) in one of my books (Chapanis, 1996). As a souvenir I received a beautiful water color rendition of the pilot boat. I had it framed and hung it on the wall of my study where it serves to remind me of a job well done.

I DESIGN A TELECONFERENCING SYSTEM

In 1980 I was engaged by Mr. Ron Carlini of IBM's General Systems Division in Atlanta to study its teleconferencing system and make recommendations for an advanced system that would make possible teleconferencing among any three of nine IBM locations. After studying the proposed system I concluded that it was much too complicated and the language much too unfriendly. In addition, control of the system was to have been through a standard computer keyboard. Various functions, sending a picture with the wall camera, sending a graphics picture, printing, and so on, were controlled by punching one of 13 keys on the keyboard. The association between functions and operations was not self-evident and had to be learned or consulted from an operator's guide.

In my March 3, 1981, report I said that "If a busy executive has to read a 30 or so page booklet in order to understand how to use a teleconference system, he is either (a) not going to use the system, or (b) use the system but curse you for it. The operation of the system should be sufficiently self-evident that a conferee could walk into the conference room and, after a brief inspection of the control panel, be able to figure out how to use it."

With those criticisms in mind I designed a control terminal with a display panel and a control panel with labels that I thought would be self-evident. I also proposed that my designs should be mocked up and tested to see whether they were in fact viable.

This was one of those unsatisfying consulting jobs. Although I visited and worked in several IBM locations in the following years I have always resented that I was never allowed to enter a teleconferencing room or to see what the final design looked like. The rooms and terminals were always "off limits" to outsiders like me.

THE WORLD CONFERENCE ON
ERGONOMICS IN COMPUTER SYSTEMS

As Robert W. Bailey, an organizer, described it, the Conference on Ergonomics in Computer Systems was too tempting to turn down. It was subsidized by a Swedish firm, Ericsson Information Systems, which paid for all travel expenses and provided generous fees for lecturers. All arrangements were to be first class. I agreed to give two lecturers at each meeting. And that was the start of one of the wildest and most exciting trips I ever made.

The Program

This is the list of speakers and what they spoke about or did:

Hakan Ledin, President and Chairman of Ericsson, Inc.: "Welcome/introductions"

John Diebold, President and Founder of The Diebold Group, Inc.: "The impact of information technology on users in the year 2000"

Robert Bailey, President of Computer Psychology, Inc.: "Overview: Ergonomics in computer systems" "Is ergonomics worth the investment?"

Patricia Wright, Scientific staff, Medical Research Council, U.K.: "Creating good user documentation"

Toni Ivergard, Manager, ERGOLAB, Stockholm, Sweden: "Workplace ergonomics: a case study"

Wilbert O. Galitz, President, Galitz, Inc., St. Charles, Illinois: "Screen design and computer messages"

Alphonse Chapanis, President, Alphonse Chapanis, Ph.D., P.A.: "Using ergonomics to increase productivity" "Rules for talking to computers"

Ahmet Cakir, Manager, Ergonomic Institute for Social and Occupational Sciences, Berlin, West Germany: "Are VDTs a health hazard?"

Tomas Berns, Director, Ergonomilaboratoriet AB - ERGOLAB, Stockholm, Sweden: "Ergonomics standards and legislation"

Brian Shackel, Head, Department of Human Sciences, Loughborough University of Technology, U.K.: "Selecting input and output devices"

Two special speakers were not ergonomists and came along for general edification and amusement. Ulf Merbold, German astronaut, spoke at some of the banquets, and James Randi, the professional magician, entertained us at banquets and on other occasions with some fascinating feats of legerdemain.

The Schedule

From September 24 to October 6, 1984, a total of 13 days, this entire assemblage convened in seven different cities at breath-taking speed:

September 24-25: The Sheraton Premier Hotel, Los Angeles
September 26-27: The Drake, Chicago
September 27-28: The Plaza, New York
October 1-2: Promenade Hotel, Amsterdam
October 2-3: The Hilton Hotel, Düsseldorf
October 3-4: Intercontinental Hotel, Helsinki
October 5-6: The Tower Hotel, London

Since the meetings in Amsterdam, Düsseldorf and Helsinki overlapped, as soon as the first day's speakers had finished, they were whisked off to the next city.

The logistics of shuttling us around must have been horrendous, but I was impressed by how well it all went considering the tight schedules. Every speaker met his or her commitment, no baggage was lost, and there were no major disasters. To a large extent it was due to the capable assistance we had. Limousines were almost always available to take us from one place to another, sometimes individually, sometimes with several of us as passengers. Helpers saw to it that our baggage got on the correct plane and that it was delivered to our hotel rooms. All our airline ticketing and hotel reservations were made for us, so that we didn't have to worry about those details.

There were a couple of close calls, though. One occasion when the scheduling was tight, two of us were ferried by helicopter from one place to another. On another occasion a flight from Düsseldorf to Helsinki was delayed several hours because of bad weather. I remember arriving in Helsinki in the wee hours of the morning, tumbling into bed only to be aroused four or five hours later to go to the lecture hall. I regretted mostly not being able to use the private sauna that came with our room. On still another occasion, five or six of us had no time to have our dinners at the hotel before boarding the small luxurious private plane that was to take us from Chicago to New York. Instead, we collected box dinners from the hotel, made off surreptitiously with several bottles of champagne, smuggled the champagne into the aircraft, and had a hilarious trip.

After the first few cities, we developed a feeling of camaraderie and a sense that we were involved in a kind of wild adventure. A number of incidents added to our sense of adventure. Our meeting in Amsterdam happened to coincide with the visit of a Mid-East king who had a suite in our hotel. Needless to say, security was very tight. Our room was on the same floor as the king's suite and we could see armed members of his bodyguard in the hallway at all hours of the day and night. In Amsterdam our hosts also arranged a gourmet thank-you banquet for the speakers and organizers in the Kastel de Wittenburg, I think it was called, all of which added to our sense of fairy-land adventure.

The Proceedings

This entire event was clearly an advertising and promotional effort by Ericsson. The newest Ericsson computers and electronic devices were on display either inside or just outside every meeting room. Our audiences were large, from about 100 to 250, and they were composed of executives, managers, engineers, and professional persons from industries and other organizations in the region.

The advertising aspect of the meetings was apparent in the 277-page *Proceedings*. Although all the technical speeches were excellent and were about the state of the art for the 1980s, they were not reproduced in the *Proceedings*. Following a table of contents, between the 8th and 9th pages of the *Proceedings*, there was a glossy 20-page brochure titled "Ergonomics and the work environment in the office," essentially extolling Ericsson's commitment to ergonomics in the office. Next were photographs of each of the speakers in turn, a brief biographical resume, an abstract of the speech, and copies of all the slides that the speaker used. I'm not quite sure why the slides, but not the texts of speeches, were included because most of these slides cannot be understood without explanations in the accompanying text. The last strange thing about the *Proceedings* is that although it lists the cities where meetings were held and the months and days when the meetings took place, nowhere does the year (1984) appear. Each speaker received a set of the *Proceedings*, but the impression one gets is that they were intended to be more like company brochures than part of the scientific and technical literature.

In Summary

It was a fantastic trip. I'm glad I didn't miss it. I met a number of prominent individuals and it exposed me to a luxurious style of travel to which I could easily become accustomed.

SAFEGUARDING AN ASSEMBLY LINE

In 1990 I was hired to make a study of worker exposure on a welder line and welder press line producing automobile parts in GM's Buick, Oldsmobile, Cadillac (BOC) plant in Kalamazoo, Michigan. My study was called Phase I of a project to study lockout and safeguarding on these lines with the ultimate goal of designing a fail-safe, automatic lockout system.

Over a period of several months I inspected production lines; studied current procedures; interviewed hourly workers, engineering and supervisory personnel; and reviewed accident records from the BOC-Kalamazoo plant.

Findings

My report, dated 18 May 1990, contained a large number of findings. To summarize them all would require more space than is warranted. Here are a few highlights:

1. There were striking differences between hourly workers and engineering personnel in their attitudes towards

safety. Hourly workers tended to think that present safeguards were adequate and that accidents were attributable to human error and carelessness. Engineering personnel, on the other hand, were much more sensitive to design for safety.

2. Current safeguards were not sufficient for many tasks and were commonly bypassed or disabled.
3. The assembly lines contained many classic human factors design faults: poorly designed dials and gauges, inadequate labeling, miswiring because of lack of coding.
4. There were serious deficiencies in the training programs for safety.

My report was accompanied by a video tape specially prepared by Jose Gutierrez showing workers performing hazardous tasks. The tape concluded with a segment showing me seated at a table summarizing my recommendations. Showing hazards and dangerous tasks visually is much more effective and convincing than verbal descriptions. The tape has been deposited with the Archives of the History of American Psychology.

EVALUATIONS OF CONTAINERS AND LABELS

In the early 90s I performed a number of evaluations of various types of containers and labels. These were performed either in my home or in my consulting offices with subjects solicited from the general consuming public. Although I think they were very good evaluations and useful to my clients, they did not involve any sophisticated human factors techniques or procedures. In all cases I used a combination of performance measures, rankings, and verbal reports. The following are reports prepared at this time:

1. "A comparative evaluation of two dispensing containers," 8 December 1992, for Specialty Packaging products, Incorporated, Richmond, Virginia.
2. "Evaluation of four sets of instructions proposed for a push-tab closure, " 26 March 1993, for Owens-Brockway Plastics and Closures, Toledo, Ohio.
3. "An evaluation of five sets of instructions proposed for a push-tab closure," 1 May 1994, for Owens-Illinois, Toledo, Ohio.
4. "A comparative evaluation of a two piece press tab closure and a conventional state-of-the-art closure," 27 December 1994, for Owens-Illinois, Toledo, Ohio.

To me, the most interesting finding was how critical it was to use the correct words in instructions. This was something I had always known (see, for example, Chapanis, 1965) of course, but these studies provided me with new evidence. For one example, we, the design engineers and I, had initially used a word "push" for a certain action. Almost all subjects interpreted this to mean a sliding forward motion. When in my tests I changed the word to "Press," everyone made the desired motion. As another example, performance tests showed that it was much clearer to say "Line up" rather than "Align."

Copies of the last three reports above have been deposited with the Archives of the History of American Psychology.

FORENSIC HUMAN FACTORS

I have been an expert witness in a rather large number of law suits, perhaps as many as 30 in all. My work was divided about evenly between testifying for plaintiffs and defendants. Although my testimony was critical in winning a substantial number of cases, I never really enjoyed this kind of work or felt a sense of accomplishment when a case was settled favorably for the client I represented. There were too many things about the process that I found disagreeable.

Antediluvian Judges

First, on a number of occasions I was not allowed to testify even though I felt the opinion I might deliver was germane to the case. I remember vividly when a judge once dismissed the jury, listened privately to my credentials and my explanation of what human factors was, and then casually dismissed me because human factors was only common sense. I felt humiliated and angry. This happened often enough to disillusion me.

The Legal Process

Another thing that disturbed me was that I thought the oath to tell "the whole truth" was a sham. In one personal injury case a worker was seriously injured while working on a construction job. I was representing the defendant, an insurance company. We had witnesses who saw the worker take lunch in a bar during which he consumed generous amounts of an alcoholic beverage. I was not allowed to testify that he was probably in a reduced state of attentiveness and dexterity because of his lunch, even though I thought this fact was part of the "whole truth" that should have been considered by the jury.

In another case in which I was testifying for the defendant, a regional telephone company, the plaintiff had a history of bringing suits against large industries over claims of injuries from rather trivial incidents. This history was not allowed by the judge even though I felt strongly that it was part of the "whole truth." As a result, the plaintiff won a substantial, and in my opinion, completely unjustified award from the company. Lawyers I have come to feel are not interested in the truth; they are only concerned with winning.

Facts Versus Verbal Skills

Furthermore, I thought the whole legal process depended too much on verbal skill. The lawyers who were most dexterous in their arguments were most likely to win. Facts were not given the weight they received in scientific work. I was never agile enough mentally to compete successfully in the verbal fencing that characterized interchanges between lawyers and experts. I recall with some embarrassment a couple of cases I'm sure I lost because I could not think of an appropriate response fast enough. It was only after I stepped down from the witness box that I could think of what I should have said.

Lawyers, Ugh!

In all my consulting the greatest difficulty I had collecting my fees was from lawyers. In fact, in one case, I had to sue a lawyer for recovery of my fees.

I was also annoyed by lawyers who claimed privilege in not revealing the amounts of monetary awards or settlements. Feedback is an important principle in psychology as well as in human factors, yet I felt that I never had proper feedback as a reward for all the effort I expended in preparing for or testifying in trials.

My experiences in this domain only reinforced the uncomplimentary regard with which lawyers are generally held.

My Deafness – A Handicap

About the middle 80s increasing deafness became a genuine handicap. In the last case in which I testified, sometime in the late 80s, I stated at the beginning of my testimony that I had a hearing loss. Despite that confession the opposing lawyer deliberately spoke in such a low voice that I had to keep asking him to repeat the question. The overall effect, I am sure, created an unfavorable impression and diminished the impact of my testimony. This was such a humiliating experience that I never accepted another case, but, to be honest, it provided me with an acceptable excuse for refusing to accept the law cases that continue to be offered me even today.

And that's all I want to put on record about this motley set of experiences.

I FACE REALITY AND QUIT

Shortly after returning from the wild and exhilarating adventures with the world conference on ergonomics in computer systems (see above), my body began to fail me. My deafness had deteriorated so much that I had to be fitted with hearing aids, but even with them I discovered that I would never again be able to hear normally. This was a distinct handicap and one that affected my ability to consult successfully.

A few months after I submitted my Phase I report on the safety project for GM (see above) management convened a conference attended by about 30 executives, managers, supervisors, and hourly workers to discuss the next phase. I was seated in the center of one of the long sides of a rectangular table and could hear only about half of what was said by various conferees close to me and almost nothing of those at the ends of the table. It was bad enough that I could no longer walk normally. Not to be able to hear normally was the last straw. It was then I knew I could not continue working on that project.

My body began disintegrating in other ways as well. From 1987 through 1996 I underwent seven major surgeries, one of them life-threatening, and twelve minor surgical interventions that were serious enough to require at least an overnight stay in a hospital. My last major surgery on January 26, 1996, required a week's stay in a hospital and an additional ten days in a convalescent and rehabilitation nursing care facility. Following my discharge from the latter on February 11, 1996, I had to accept the bitter reality that I would never again be able to walk unaided, would never again be able to hear normally, and that I would never have the vigor I had even in my 70th year.

All this meant that I could never again be an effective consultant. Even lecturing was excluded not only because standing for an hour was painfully excruciating but also because I could not engage in meaningful interchanges with my listeners. And so, regretfully and sadly, I closed my comfortable consulting offices in

November 1996. One of the few intellectual activities still open to me is the activity in which I am now engaged – writing.

12. EDUCATING BIG BLUE
1959-1995

Of all the organizations and institutions with which I have been associated, the one for which I have the greatest regard (I am almost inclined to say affection) is IBM. My association with this company spans a period of 36 years, which, coincidentally, is exactly the number of years I was officially associated with The Johns Hopkins University.

In Chapter 6 I describe research studies that IBM supported in my laboratory while I was still a professor, and in Chapter 11 I described a number of studies I performed for IBM. In this chapter, however, I am more concerned with educational activities. I think I can say without exaggeration that I played a prominent role in educating thousands of IBM executives, managers, engineers, and programmers about human factors. There were, of course, many capable and talented human factors professionals in IBM, but they were generally much too involved in their various activities to have time to devote to education. I, on the other hand, came to IBM as an outsider, an educator, primarily to teach. In so doing I think I contributed an outsider's viewpoint and broader perspectives about human factors than did the professionals within IBM whose interests were generally closely tied to their work.

This was not a one-way street, however, because I learned more *from* IBM than IBM learned from me. Working closely with executives, engineers, programmers, and researchers of all kinds was a unique educational experience. Even more important was that I learned how human factors was used and how it should be used in the design of machines, equipment, and systems. After having had that experience I would teach human factors much differently in the university if I had the chance to do it over again.

FOR THE OFFICE PRODUCTS DIVISION

Unfortunately, I have no documentation about my work up to 1980. So for the period prior to that time I have had to rely on fragmentary notes and some recollections from John Shafer to augment my own. I know that the first real course I taught for IBM was for its Office Products Division in Lexington, Kentucky, probably in the early 70s. It was a good course and I remember preparing some excellent illustrations about human factors faults I could find in the design of typewriters and what human factors could contribute to the design process. I used these illustrations over and over again in later lectures.

AT IBM'S RESEARCH CENTER

Actually, my long-time association with IBM began in 1959 when I spent the summer at its research center then in the Lamb

Estate, Ossining, New York. This was before IBM built its elegant new laboratory in Yorktown Heights. I don't remember much about what I did there, but I do recall that Edmund T. Klemmer was director of human factors and that Nancy S. Anderson and Gustave Rath were also there. Among other things, we collectively spent some time outlining future directions the research center should take.

AT IBM'S ADVANCED SYSTEMS DEVELOPMENT DIVISION

After my return from Europe in 1961 (Chapter 9) I was for several years a consultant to what was then the Advanced Systems Development Division (no longer in existence) in Yorktown, New York. I typically spent about two days a month there advising researchers about various projects.

On the Importance of Selecting the Right Subjects

I have used one case study from ASDD in my lectures on research methodology. Someone, I don't remember who and it really doesn't matter, had developed a computer program for recognizing handprinted numerals. At that time all this kind of work was in its infancy. Tests showed that the program successfully recognized over 90% of handprinted numerals submitted to it. When I studied the protocols, however, I discovered that the test subjects had all been not only IBM employees, but IBM accountants as well. At my strong urging the study was repeated using samples of numerals printed by people in the neighboring town, and in these trials handprint recognition dropped to about 20%. An impressive example of the importance of choosing test subjects appropriately.

On the Importance of Understanding Computer Programs

A second incident also provided me with additional lecture material. One of the researchers, who I shall leave nameless to avoid embarrassment, had analyzed a set of data with a computer analysis of variance program. When I examined the experimental design I determined that the sources of variance and F-tests could not possibly be legitimate ones. It turns out that this researcher had selected an inappropriate analysis of variance program, had keypunched in the raw data, and accepted uncritically the results printed out by the computer. When we together reanalyzed the data using the correct analysis of variance model, an entirely different set of significances emerged.

I have often used this as a moral for my students: You can't accept computerized analyses uncritically just because they're done with computer programs; you have to know and understand your data and what the computer is doing with them.

I Acquire a Graduate Student

One of the assistants working at ASDD was a young man named Robert N. Parrish. I was much impressed with him and urged him to apply to graduate school in my program. He did, and received his Ph.D with me in 1973.

ADVISOR TO CORPORATE RESEARCH

Shortly after my ASDD consultancy I was for a time an advisor to Lewis M. Branscomb, then a vice-president and chief scientist at IBM's Headquarters in Armonk, New York. It was then that I recommended the program that follows.

A CORPORATE HUMAN FACTORS COURSE

According to John Shafer's notes my involvement in this program started in June 5, 1978, when Richard Pew and I were invited to attend a planning session at IBM's corporate headquarters. The meeting was attended by at least three high-level IBM executives, Lewis Brascomb, Jerold Haddad, and B. O. Evans; by human factors personnel from various IBM divisions, John B. Shafer, Charles R. Brown, Ronald Beechler, Lance A. Miller, Irwin Miller, Richard S. Hirsch; and by perhaps other persons. One of the decisions made at that meeting was to have a Human Factors Symposium that would make a large number of key IBMers aware of human factors.

The Awareness Symposium

The symposium was held early in 1979 in Denver (Brown's Hotel, as I recall), and I was the keynote speaker. There were about 250 key IBM people in attendance. I gave what was generally regarded as a good introduction to human factors. By then I had accumulated a sizable number of examples of poor human factors in computer systems and I wove these into my speech, bringing forth embarrassed laughter and groans of recognition.

A videotape of my speech has been deposited with the Archives of the History of American Psychology. The tape only bears the title "Human Factors," but a letter of transmittal from

IBM accompanies the tape.

A Planning Meeting

Shortly thereafter, on March 8, 1979, according to Shafer, there was a meeting in Haddad's conference room in White Plains, New York, with Robert DeSio, Stephen Komjathy, Ben Blau, Dave Seddon, Olaf Henkle, Claude Erland, Shafer, and I attending. I remember that meeting well, because it was at that time that a decision was made to develop a human factors course that would be taught at key IBM sites around the world. I was designated the consultant to be in charge of the development. I protested, however, that although I knew a lot about human factors I didn't know very much about what went on in IBM. For that reason I needed some very knowledgeable person from within IBM to assist me. Shafer was designated to be that person and, as he said in his email message to me, "It was a whirlwind from then until July 4, 1980" when the course was last taught. And indeed it was!

Preparing the Course

For the next several months after the March 8 meeting I spent two days a week at the IBM building in White Plains, New York, planning, organizing, and writing. Personnel from every IBM human factors department were brought in, singly or a few at a time, to contribute their ideas, data, and examples of accomplishments. The list was extensive and included at least the following: Lou Adams, Doug Antonelli, Ron Beechler, Paul Franks, Art Gershon, Dick Granda, Ralph Grubb, Dick Hirsch, Bill Hrapchak, John Hughes, Jim Judisch, Pete Kennedy, Win Miller, Lance Miller, Mark Mullen,

Mark Ominsky, Dave Peterson, Hans Plotzeneder, Phyllis Reisner, Dave Seddon, Tony Silvestro, and Paul Tobias. Ben Shneiderman from outside IBM was also brought in for some ideas about the application of human factors in computer systems. Jerry Hoffman served as a sort of librarian, collecting and cataloging all the materials as they accumulated, a vital organizational function. Since Shafer was still based on Owego and came to White Plains the same two days a week as I, Ben Blau became the local person in charge.

The pace was frenetic, but the working environment heady. IBM allocated almost unlimited resources to this venture, and nothing that I wanted or requested was denied for lack of funds. For example, I wanted a filmed short introduction to each lecture, a kind of prelude to what was to follow. One of these has always delighted me because it was so effective. In my laboratory at CRL we had been using a programming language called APL, and a peculiarity of that language gave me an idea for an introduction to the software lecture.

An effective opening. The film opened with a man standing in front of a blackboard. He said something like, "In this country we speak English and write..." and with that he turned to the blackboard and wrote: sin X, cos X, tan X. That scene dissolved and was replaced by a woman speaking similar lines in German at the conclusion of which she turned to the blackboard and wrote under the first set: sin X, cos X, tan X. That scene dissolved and was replaced by a man speaking Chinese and writing under the first two lines: sin X, cos X, tan X. The final scene showed a man who said something like, "In computer programming we use APL and write..." after which he turned to the blackboard and wrote: 1oX, 2oX, 3oX.

That little opener invariable resulted in guffaws. It was so successful that I had a set of slides made up to convey the same message and used those slides innumerable times. They are illustrated in the talk I gave at the ITT symposium in 1982 (Chapanis, 1982c. p. 46).

Organization of the Course

As it finally evolved the first day of the course consisted of (1) an introduction to the program by the local IBM site manager, (2) a film by B. O. Evans, an IBM executive, on the importance of human factors, (3) a lecture on human factors as a profession, (4) a film by R. DeSio, another IBM executive, on the relevance of human factors to IBM, (5) a lecture on the European scene and the impact of European standards on IBM business, (6) a lecture on human factors test and evaluation methods, and (7) a lecture on human factors in user requirements.

The second day of the course continued with (8) a lecture on human factors in the development process, (9) a lecture on human factors in product installability, (10) a lecture on human factors in documentation, (11) a lecture on human factors in maintainability, (12) a lecture on human factors in software, (13) a film on forecasting and human factors, and (14) an open session for comments and discussion.

Another effective video tape. For the lecture on methods we had a video tape illustrating the use of stop-action or lapse-time photography in activity analysis. One segment showed difficulties some employees had in using an IBM time and attendance recorder. One employee was so confused he had to call for assistance. Another segment showed difficulties an IBM employee had loading paper into an IBM printer while reading the instruction manual.

173

This was an especially impressive segment because it showed the employee going from front to back repeatedly and eventually calling on a supervisor for assistance. The voice-over concluded with something like, "The manual is so confusing that even an IBM employee can't figure out how to load the printer." The last segment showed how eight hours of office activities could be usefully captured with about 5 minutes of recording.

This video tape was extremely useful as a teaching device. One can talk about these methods and their use of revealing human-machine difficulties, but actually seeing them in practice helps reinforce the spoken words. One other video was used for the lecture on human factors in maintainability. It was an action sequence without words. It showed an IBM service representative trying to service an automated bank teller machine. The service man made several attempts to worm his way up into the back end of the machine and in so doing he was repeatedly bumped by parts of the machine that eventually tore his shirt. Another impressive film that spoke for itself.

Teaching the Course

The course was completed sometime during the fall of 1979 and Shafer, Blau and I taught it several times in 1979. The only documented date I have is for the course that we taught for IBM Hursley in the Hotel Potters Heron, Ampfield, England, on October 10 and 11. Starting early in 1980 the course was taught by Shafer and Blau (I could no longer participate because it would have meant taking a leave-of-absence from teaching at Hopkins) 15 times at various IBM sites in the United States and 10 times in Canada, France, Germany and Japan. The last session was on July 4, 1980, in

Tokyo.

A SOFTWARE AND INFORMATION
USABILITY SYMPOSIUM

Some time in 1980 I was invited by Terry Gair in IBM's System Communication Division, Kingston, New York, to organize a symposium on human factors as it related to software and information usability. My job was to select a panel of speakers who could present the state of the art on usability outside of IBM. I was given freedom to select the participants, invite them to the symposium, and edit their manuscripts. Speakers were, of course, to be well paid by IBM. One of my former students, Charles Brown, now an IBM employee at Kingston, New York, gave formal IBM approval of my selections.

Content of the Symposium

The symposium was held on September 15-18, 1981, in the IBM 701 Auditorium in Poughkeepsie, New York, with about 300 attendees. Those of us from outside IBM were only allowed to attend the first day. For that session I had enlisted Brian Shackel (The University of Technology, Loughborough, U.K.) who spoke on "The concept of usability;" Sidney L. Smith (The MITRE Corporation, Bedford, Massachusetts) who spoke on "The usability of software – design guidelines for the user-system interface;" and Patricia Wright (MRC Applied Psychology Unit, Cambridge, England) who spoke on "Problems to be solved when creating usable documents." I spoke on "Evaluating ease of use." Our papers were published in Volume 1 of the *Proceedings*. Although the volume was labeled "IBM Internal Use Only," we were each given a copy of that volume. We were not, of course, provided with copies of Volumes 2 and 3, which dealt with evaluations of the usability of current IBM SCD products and usability activities throughout IBM.

Figure 8. Speakers at the IBM Software and Information Usability Symposium September 15, 1981. From left to right: Alphonse Chapanis, Patricia Wright, Terry Gair (IBM Symposium Manager), Sidney Smith, Brian Shackel.

I gave a slightly altered version of my IBM talk as an invited address at the Advanced Study Course held at Loughborough University, England, on December 14-19, 1986 (Chapanis, 1991d). The course was sponsored by the European Science and Technology Research Committee of the Commission of European Communities and by the Science and Engineering Research Council of the United

Kingdom.

FOR IBM'S QUALITY INSTITUTE

From about 1982 to 1990 I gave a four-hour lecture on human factors for IBM's Quality Institute as many as eight times a year. The lectures were part of a week-long course designed to train selected IBM employees to become quality control engineers. For the first few years the lectures were held in the Hilton Hotel, Danbury, Connecticut, where IBM had rented some space. Beginning in 1985 or 1986, the course shifted to IBM's elegant new Education Center in Thornwood, New York. One course, on February 2, 1988, was given at an IBM site in Los Gatos, California. It was cheaper to have the instructors travel to California than to have the students come to the East Coast. There was nothing really memorable about these lectures; they were introductory lectures that made use of material I borrowed from the corporate course described above.

IBM'S FEDERAL SYSTEMS DIVISION

In March 1982 I gave an invited address at a conference on "Human Factors in Computer Systems" held in Gaithersburg, Maryland. The meeting was sponsored by the Institute for Computer Sciences and Technology, National Bureau of Standards, Washington, D.C. and the Washington, D.C. Chapter of the Association for Computing Machinery. It also had the cooperation of the Office of Naval Research, Engineering Psychology Programs, the Software Psychology Society, Computer Systems Group of the Human Factors Society, and IFIP Working Group 6.3.

Unfortunately, I have no copy of my address and it was not

printed in the *Proceedings* of the conference, but it must have been a good speech because after my talk I was approached by Harlan Mills, an IBM Fellow, who had been in the audience. Mills wanted to know whether I would be interested in becoming a consultant to what was then the Federal Systems Division (FSD) of IBM with headquarters at 6600 Rockledge Drive, Bethesda, Maryland. I said I would be, and thus began a long, interesting and productive association that lasted until 1995.

FSD as a Division

FSD was perhaps the most interesting division of IBM in which to work. Although it was a small division producing income of only a few billion dollars a year, it was always a profitable division. Unlike some other divisions it never lost money. It also had the highest proportion of engineers and programmers of any IBM division because it was responsible for the design and development of large military systems, LAMPS, B-52; large government systems, automated post office, air traffic control; and large commercial systems, automated banking. I have never been able to understand why IBM eventually sold FSD to LORAL Federal Systems in 1994. But I digress.

PROFS. One of my first assignments was to critique a system called PROFS, Professional Office System. This was a new IBM system that enabled users to communicate with one another, keep diaries, store and recall reports and documents, and assist in sundry other office tasks. PROFS terminals had been installed throughout FSD's headquarters in Bethesda, Maryland, but were meeting with resistance from the engineers, managers and other professional persons. The question put to me was "Why?" I was given an ID to access the system and a terminal on which to work. The user's guide, incidentally, stated that the system is for "those who have little or no experience with computer systems."

Suffice it to say, I quickly discovered a number of human factors deficiencies. Although I no longer have a copy of the critique I wrote, I have used and documented several of my findings in talks and articles, for example Chapanis (1984). My first criticism was that the display was entirely in upper case letters, and I recommended that the format be in upper and lower case. Time was kept in Julian days and I took great pleasure in saying sarcastically, "for God's sake, what ordinary person knows or cares about Julian days?" Accessing the system required an extraordinary number of steps (16) which I reproduced faithfully on page 220 of the Proceedings of the 1984 World Conference on Ergonomics in Computer Systems (see Chapter 11). The system was also rife with arbitrary and confusing abbreviations and commands and I disparaged the 12-digit document codes, such as 82252TST0002 and 81231HDC0019, with no gaps or dashes to make their use and recall easier. There were other problems too, but I can no longer recall what they were. Suffice it to say my report bore fruit, and the PROFS system that finally evolved is a much more user-friendly and useful system.

Systems Engineering Principles and Practices

In the early 1980s a small group of engineers (William J. Budurka, Frank Testa, and Robert Wheeler) was working to systematize the practice of systems engineering. As part of that effort, they developed a week-long course called "Systems Engineering Principles and Practices" (SEPP). One small part of that course was a two-hour segment on "Human Factors" prepared by an IBM employee. One day they invited me to review that segment. I criticized it severely because it was essentially an academic discussion of human factors, but didn't tell the engineers how to use or apply human factors in system development.

My criticisms apparently agreed with their own evaluations because they then invited me to prepare a better segment. As I did with the corporate course a few years earlier, I requested and received approval to have John Shafer work with me to provide me with his experience in applying human factors in the development of systems. The new version of that material was inserted into the students' books in March 1984, and my notes indicate that I first taught it on March 28, 1984. Student evaluations gave it high marks and it continued to be rated high in courses taught later that year in Bethesda and Owego, New York.

Human Factors in Systems Engineering

The continued success of the human factors segment in the SEPP course encouraged Shafer and me to suggest that we could prepare a much more thorough course on the same subject. With the support of our engineers, we received approval from our manager to proceed. The first version of "Human Factors in Systems Engineering" (HFISE) was completed in June 1985, and Shafer and I taught it for the first time that year. As we continued to teach it, we kept seeing ways in which it could be improved. As a result, the course went through nine successive revisions, the last in June 1995.

The student books are 434 pages in length and consist of eight major chapters: introduction; standards, specifications, and reviews; human factors methods; human characteristics; interface design; personnel selection and training; principles and practices exercise; and summary. The course has some unique features. First, it is not an academic presentation of the contents (or nature or essence) of human factors, but is rather oriented around the systems engineering process, that is, what human factors can contribute to each step of the design and development of systems. Our eyes were always focused on the question, "What does the systems engineer need at this time and in what form should human factors supply that information?" This orientation was the result of close monitoring by the engineers, who continually asked such questions as "What does that tell me, an engineer, that I can use?" "What does that contribute to the design and development of a system?"

A second major feature of the course is that it emphasizes the importance of standards, specifications, and reviews – something that is totally absent in all human factors textbooks. In actual practice the systems engineer is constrained by specifications and standards that are imposed on him by the government, by the customer, or by his own industry. Moreover, systems engineers produce more words than they do things. For a large system they typically have to prepare an operational need document, an operational concept document, specifications for the system as a whole and for each of the major subsystems and components, requirements rationale reports, design descriptions, and design rationale reports. All these work products contain, or should contain, human factors requirements, and in our course we show where and in what form these human factors requirements should be stated. Throughout the course we show how this was done with actual systems that were developed by FSD.

Exercises. Another feature of our course is the liberal use of exercises. Each exercise is designed to be short enough to be completed in no longer than a half hour, to demonstrate some important human factors principle, and to be inherently interesting. Our exercises have always received highly favorable student evaluations. The last four hours of the second day are devoted to a longer practical exercise in which the students have to use much of the material they had been taught in the preceding day and a half. For the course we also used the video tape on lapse-time photography in activity analysis and the tape on maintainability that we had produced for the corporate course.

Shafer and I taught the course together, alternating in our lecturing, about 30 times according to my count: in Bethesda, Rockville and Gaithersburg, Maryland; Owego, New York; Manassas, Virginia; Houston, Texas; Boulder and Colorado Springs, Colorado; and Sunnyvale, California. When IBM sold FSD to LORAL Federal systems in 1994, I taught it for the last time with Shafer under LORAL's auspices. Since LORAL did not want the course taught by outsiders, I was replaced by Lou Adams, and my association with IBM and LORAL ended. I accepted these moves regretfully but without resentment because it was time for me to quit teaching. My physical condition and my deafness in particular had progressed to the point where I could no longer interact with students, thereby greatly reducing my effectiveness. Shafer and Adams continued to teach it for LORAL, and at the time of this writing they continue to teach it for Lockheed Martin after LORAL sold out to Lockheed.

The course was always labeled "IBM [or LFS] internal use only" and neither Shafer's name nor mine ever appeared on or in the books. I later used much of this course material in my book (Chapanis, 1996b) and for 10 years Shafer and I gave workshops on human factors methods at annual meetings of the Human Factors and Ergonomics Society based on material from the HFISE course. We also wrote an article describing the general nature and contents of the course (Chapanis & Shafer, 1986).

During the preparation of this course I developed a profound respect for engineers and their profession. They, and Budurka in particular, were sharp, witty, imaginative, and articulate. I learned more from them than I ever taught them.

A Human-Computer Interface
Requirements Specification (HCIRS)

While we were developing the HFISE course the engineers would often disparage human factors guidelines that were, and still are, being published in great profusion. "They're not specific. They're so general they can be interpreted in different ways and leave the choice of alternatives to me, but I'm not a human factors professional." Comments such as these led me to make the brash assertion that I could write a set of instructions for a project-specific human-computer specification that was a true specification and not a set of guidelines.

So sometime around 1985 Budurka said to me, "OK, show me." Given that challenge I struggled for months, but eventually produced a draft of what I thought would be such a document. Shafer reviewed it critically and added his own contributions. Budurka also had his own comments about it. The document went through several such iterations among the three of us before it was distributed widely throughout IBM for comments from programmers, engineers, and human factors professionals. That produced still more comments. But in the end I was immensely pleased when it was formally adopted as an IBM Corporate

Bulletin (Number C-B 0-2507-011) under the title "Federal Systems, Human-Computer Interface Requirements," and included in IBM's Federal Systems Division Manual 10-09, "Systems Engineering Standards."

When IBM's FSD was purchased by LORAL Federal Systems, LORAL adopted the HCIRS verbatim, and when LORAL was purchased by Lockheed Martin (LM) it also adopted the HCIRS but with a new designation. It is now "Systems Engineering: Human-Computer Interface Requirements (HCIRS) Number GB 0-2057-011" in LM's Software & Systems Resource Center Process Guidance Series. In any case, the HCIRS was apparently good enough to have withstood two changes of corporate ownership.

As usual, the HCIRS is marked "Internal use only" and neither my name nor Budurka's nor Shafer's appears on the IBM or LORAL versions of the HCIRS. We were, however, given permission by IBM to write an article describing its general features (Chapanis & Budurka, 1990). Ironically, the LM version lists Joe DeFoe, an LM employee, as editor.

I'm proud of the HCIRS. I think it's one of the best things I ever created. For the first time, I was able to articulate clearly exactly what human factors professionals should be doing. Rather than quoting or citing general guidelines or recommendations to engineers, they should be writing project-specific human-system, human-machine, human-computer, or human-device specifications. I tried to explain this philosophy in Chapter 8 in my 1996 book, but, as I have discovered when I have lectured about it, this point of view is difficult for many human factors professionals to understand and accept. My only regret is that I was never allowed to see how the HCIRS was actually used in the development of some system. I am certain, however, that it had some impact in making computer systems easier to use.

IN RETROSPECT

In looking back at my association with IBM as I have described it in Chapters 6, 11, and this one, I have feelings of satisfaction at my accomplishments. I did some good research with and for IBM, I taught thousands of IBMers about human factors, I designed or contributed to the design of some successful IBM products, and I enjoyed doing it all. What more could a human factors guy ask for?

13. LECTURESHIPS
1971-1996

I have already written about a large number of talks I gave during the Systems Research period (see Chapter 3). In the years following I was also invited to give a number of lectures for diverse organizations and to visit a number of colleges and universities as a visiting lecturer, scholar, or professor. Some were evidently uneventful because I have little or no recollection about what I did or said on those occasions. Among them were lectures at Williams College, Williamstown, Massachusetts, April 1972; Union College, Schenectady, New York, April 1972; Iowa State University, Ames, Iowa, January 1975; Virginia Polytechnic Institute and State University, Blacksburg, Virginia, February 1976 and April 1977; University of Houston, Houston, Texas, October 1976; Stevens Institute of Technology, Hoboken, New Jersey, April 1976; National Telecommunications Conference, New Orleans, Louisiana, December 1976; Xerox Corporation, Los Angeles, California, May 1977; Texas Instruments, Dallas, Texas, July 1977; Committee on

Telecommunications, Interagency Committee in Telecommunications Research, Washington, D.C., October 1977; IEEE National Telecommunications Conference, Dallas, Texas, November 1977; Systems Research Institute, IBM, New York City, February 1979; International Seminar on "Perspectives in Ergonomics", Istituto per la Formazione e l'Aggiornmento Professionale, Rome, Italy, 1981; International Information Industry Conference, Quebec, Canada, June 1-2, 1982; and the International Course on Human Factors and Ergonomics organized by the Nordic Institute of Advanced Occupational Environment Studies, 17-21 October, 1983.

I remember a number of others, but regard them as being either rather ordinary or not worth documenting. Among them are General Foods Corporation, August 14, 1985; MITRE Corporation, McLean, Virginia, January 21, 1987; and the Consultants Division Seminar at the ASSE Annual Development Conference, Baltimore, Maryland, June 13, 1987.

The following lectures and visits were, for diverse reasons, more memorable than most.

A NATO LECTURESHIP

According to an article in *The Johns Hopkins Gazette* (December 11, 1971) I was the 1971 speaker at universities and research organizations in five countries under a visiting lectureship program of the North Atlantic Treaty Organization. According to the article I lectured on my communication research in Great Britain, France, the Netherlands, Germany, and Norway. Two of these visits stand out in my memory.

A Visit with Murrell

In Britain I lectured at the University of Wales in Cardiff. The ergonomics laboratory was then under the direction of Hywel Murrell (now deceased), the man who had coined the term ergonomics. After my talk he drove me out to his sheep farm for a delightful dinner in his unique fifteenth-century home with saddles, bridles, and other farm equipment hanging from the ceiling. Hywel gave me a fascinating account of how he had traced the history of his home through old documents.

At the Forschungsinstitut für Anthropotechnik

My German lecture was at Rainer Bernotat's Forschungsinstitut für Anthropotechnik in Meckenheim. The Institute had a rather small staff and Bernotat had evidently prevailed upon some persons who did not speak English to attend my lecture. When he introduced me in English a few members of the audience settled back in their chairs with bored expressions on their faces. But when I started to lecture in German they suddenly perked up. Before I left home I had taken the precaution of having my lecture translated into flawless German and practicing my delivery with the help of a German professor at Hopkins.

I REVISIT ISRAEL

According to a newspaper article in *The Evening Sun* (December 29, 1971) I visited Israel in December 1971 and gave lectures on my interactive communication research at the Eliezar Kaplan School of Economics and Social Sciences of the Hebrew Institute in

Jerusalem and at the Technion, the Israel Institute of Technology in Haifa. Haifa is not far from the Lebanon-Israeli border and I had the unusual experience of listening to the muffled sound of cannon-fire while lecturing there.

I was received cordially both in Jerusalem and Haifa and after my lectures spent hours talking informally with staff and students who seemed to be hungry for information about the US and what I did. But the thing I remember most vividly about that trip was a mishap that many people have experienced.

I am a Victim of a Common Airline Blunder

I was traveling around the world and had spent some time in Australia. From there I flew to Singapore for an overnight stay. In leaving Singapore I checked my baggage at the airport and that was the last I saw of it until I had returned to Baltimore a couple of weeks later. I was traveling in casual clothes, but fortunately had kept my passport, travelers cheques, and, most important, my slides with me in an attaché case. After confirming in Tel Aviv that my luggage was really lost, TWA allowed me to purchase an extra pair of pants, some other basic clothes, and shaving materials. For the rest of my stay in Israel, I lectured and traveled in casual clothes with my belongings in a paper bag and my genuine essentials in my attaché case.

Despite this inconvenience, I had a successful and interesting time. I had rented a car and saw the Dead Sea, climbed to the remains of the ancient fortress at Masada, and traveled through the newly acquired but hostile West Bank. In Jerusalem I stayed overnight at the Intercontinental Hotel, acquired by Israel from Jordan in the 1948 war. Although officially on Israeli territory, the

189

hotel was managed according to Arab customs. The entire staff, including the "chambermaids," was male. My baggage, I found out later, had ended up in Zürich and was delivered to my home a couple of weeks after my return.

A VISIT TO RUMANIA

The Johns Hopkins Gazette for September 26, 1974, carried an article about my being invited to address the International Symposium on Practical Applications of Ergonomics to Industry, Agriculture, and Forestry in Bucharest, Rumania, September, 1974, as a guest of the International Labour Office, Geneva, and the Rumanian Ministry of Labor. According to the article the title of my talk was "What does ergonomics have to do with worker satisfaction?" I remember making such a talk and I wish that I had published it or at least kept a record of what I said, but, unfortunately, I didn't.

VISITING PROFESSORSHIP –
UNIVERSITY OF MISSOURI-ROLLA

An article in the February 8, 1980 issue of the *UMR Digest*, states that I was a visiting professor at the University of Missouri-Rolla during the week of February 10-16, 1980. In addition to conferring with students about their research I gave two public lectures. One was on "The Importance of Engineering Psychology in a Technological Society" on February 11 (*UMR Digest*, February 8, 1980, p. 1), the other on "The Human Error at Three-Mile Island" on February 15 (*Rolla Daily News*, February 17, 1980).

The lecture on Three-Mile Island was based on an article that I had written and never had published. It was a good well-written article, editors of the *Scientific American*, *American Psychologist*, and a couple of other publications agreed, but it was too late. So many articles had already been written about that catastrophe that no editor was willing to accept mine. Still, it was not a total loss because I was able to use it on several other occasions.

THE UNIVERSITY OF MICHIGAN

Paul Fitts was, I believe, the first to organize the University of Michigan summer short course on Human Factors Engineering. After his death it was taken over by Richard Pew and Paul Green and at the time of this writing they were still directing it.

This two-week course is given every year and without question is the best of the many short courses on human engineering that have sprung up in the intervening years. It was, and is still, I'm sure, immensely popular. Attendances hover around 100 and the audiences have been made up of engineers, designers, managers, professors, and students from all walks of life. From about 1975 to 1985 I gave three lectures each year at these courses: basic vision, visual displays, and methodology.

I had and still have mixed feelings about these courses. On the one hand, I think they serve a useful purpose in introducing a large number of people to human engineering. On the other hand, I am afraid that many attendees go away with the feeling that after having attended this two-week course they were now qualified human factors professionals. And, of course, no two-week course is enough to do that. I conveyed this concern to Pew on several occasions. He was in agreement with me and said he tried to get this message across to the attendees. Still, on occasion I have seen

191

individuals claim expertise in human factors when the only training they had received was this course.

I taught in this course until 1985 when a trip (I planned to attend the IEA meeting in Australia) conflicted with the dates of the course that summer. After ten or so years my lectures were becoming a little stale and boring, so it was a good time for me to quit, despite the laudatory letters I used to receive after these lectures. Following is an excerpt from a letter written to me on August 27, 1985 from a David B. Stover of Martin Marietta Baltimore Aerospace in Bel Air, Maryland:

> I would like to reiterate what an honor and pleasure it has been making your acquaintance and at least being your "part time" student at the University of Michigan - the whole experience was fantastic! I do regret that I could not be one of your full time students, in the usual academic sense of the term; however, I intend to be one of your "students" as opportunities arise.

INDIANA UNIVERSITY

For about three years, from 1976 to 1979, I gave lectures at a course organized at the University's Naval Safety School in Bloomington, Indiana. The course was given for Navy civil servants and engineers and was intended to teach them about the latest in safety. My lectures were on information theory. Although I tried as best I could to point out some practical applications of information theory, I could never understand why the organizer thought safety engineers should know about this subject. My lectures were apparently satisfactory since I was invited back several times until

1979 when I decided the honorarium they were paying me was not worth the trouble.

A SYMPOSIUM ON UNIVERSITY CURRICULA
IN ERGONOMICS

From 11-15 March 1976, I participated in a symposium on *University Curricula in Ergonomics* held in Berchtesgaden, West Germany. According to *The Johns Hopkins Gazette* for April 15, 1976, I delivered the opening address (which again I have no record of) and chaired a working session on the preparation of a curriculum based on the views of all the participants. This eventually resulted in a booklet (Bernotat & Hunt, 1977) which did not receive much attention in the US. A photograph of the 16 participants is on page 4 of the booklet.

An Excursion into Austria

For this meeting I had rented a car (a Mercedes-Benz). One of the participants, Professor Dr. Carl Graf Hoyos, said he knew an excellent restaurant in a little village in Austria not far from the German-Austrian border. I offered to drive him and two or three other persons to the restaurant. At the border I was challenged by the German border guard who thought I might be trying to abscond with the automobile. I got tangled up in my German verbs using *leihen*, meaning "to loan" when I meant *mieten*, meaning "to rent." The guard was giving me a hard time, when Hoyos came to my rescue. He showed the guard his identification card whereupon the guard promptly saluted and waved us on. Hoyos is a Count, legitimate royalty. *Graf* in German means "Count." At the restaurant we had some fine Austrian wine and an excellent meal during which I was teased unmercifully about my faulty command of German.

UNIVERSITY OF LEUVEN

In 1977 Professor Paul Verhaegen at the Catholic University of Leuven, Belgium, invited me to spend the month of May 1978 as a Distinguished Visiting Professor in the Institute of Psychology at the University. According to a press release I gave some general lectures on human factors engineering to students at the Institute, emphasizing American approaches, concepts, and ideas. I also gave a general University lecture on my research in telecommunications and teleconferencing. My teaching schedule was sufficiently light to permit me to travel and give lectures on my research at the Instituut voor Perceptie Onderzoek, Endhoven, The Netherlands, and at the Department of Psychology, Johnannes Gutenberg University, Mainz, Germany. I also made a professional visit to the IBM Research Laboratories at Hursley Park, UK.

My Accommodations

At Leuven I had an apartment in the *Groot Begijnhof*, a restored and renovated nunnery dating back to 1628. As I described it in a letter to my secretary, my apartment had a large living room with high beamed ceilings, a fireplace one could walk into, and shuttered windows with small panes of hand-blown glass. It was, however, furnished with a modern kitchen and other modern conveniences. Narrow, cobble stoned streets and two branches of a river meandered through the apartment complex. It was the most

194

picturesque place I have ever lived in.

A PROPHETIC LECTURE

In July 1977 I gave the first human factors lecture at Babcock and Wilcox, Naval Nuclear Fuel Division, Lynchburg, Virginia. This firm was largely responsible for the design and construction of the Three-Mile Island Nuclear Power Plant. If my talk did not persuade them to hire a human factors specialist, the accident at the plant in 1979 most certainly did.

THE UNIVERSITY OF SOUTH DAKOTA

In the spring of 1991 I spent 6 weeks at the University of South Dakota, Vermillion, South Dakota, as a Distinguished Visiting Professor. At the time I had drafted several chapters for my forthcoming book (Chapanis, 1996b) and I used them as the basis for a graduate course I taught during my stay there.

I also gave a public lecture on "The Human Error at Three-Mile Island" – the same one I had given at the University of Missouri several years before (see p. 190). The lecture was received enthusiastically providing evidence to me that it was a good lecture and had not lost its human interest.

THE UNIVERSITY OF CONNECTICUT

My last visiting professorship was at the University of Connecticut in September, 1996. It was a visit of nostalgia because I had graduated from that institution in 1937 when it was still Connecticut State College. Seeing the dormitory where I had lived, the Armory in which I had trained as an ROTC cadet, the swimming pool in which I had swum as a member of the swimming team, the small river in which we swam when the weather was warm, and sampling the gourmet ice cream manufactured from the University's own dairy brought back memories of some of the happy times I had spent there as an undergraduate. This visit was also my last in a literal sense. During one of my 50-minute lectures, my legs pained me so much that I had to be helped away from the lectern. I knew that I could never accept an invitation to do something like this again. It was a shattering realization!

Some of My Activities

I participated in a number of seminars taught by my genial host, Don Tepas, and informal "brown bag" luncheons the department held once a week. Informal advisory meetings with Rob Henning's graduate students reminded me of how much I used to enjoy similar sessions with my own students at Hopkins. From time to time students would drop into my office for advice, or simply to chat.

In one of the seminars I taught for graduate students I used all the exercises John Shafer and I had devised for IBM's Human Factors in Systems Engineering course (see p. 181). I told the students that after their graduation they might have to have the job of convincing skeptics about the value of human factors and that these exercises might help them to do that. The exercises were not only short, interesting, and fool-proof, but each one demonstrated some important human factors principle. I had also worked up lists of relevant bibliographic articles they could use as reference material for each exercise. The session went over very well and reinforced what I had already known – the exercises were indeed interesting and informative.

On several occasions I thought that I should write these exercises up into a paper bound manual for sale to teachers and instructors of human factors. I'm sure they would be useful as teaching aids. Another job for me in my retirement years!

Two Successful Lectures

When one receives a Distinguished Contribution for Applications in Psychology Award from the APA, he or she is expected to deliver an address at the next annual meeting and to have the address published in the *American Psychologist*. As required by custom, my address was delivered on 4 September 1979 at the annual APA meeting held in new York City. It went over well, as was apparent from responses I received from listeners. For example, in one letter Rufus C. Browning wrote:

> Your paper "What the Practical World has Taught me about Basic Research" was, in my opinion, the best presentation at this year's A.P.A. I would appreciate having a copy of the paper for my files. Further, I hope you plan to submit the paper to the *American Psychologist.* Your paper deserves far more circulation than that provided by an A.P.A. live audience.

The lecture I gave on 25 September 1996 at UConn was the same one I had given in New York City in 1979. At UConn it was received with laughter at times, nods of the head at other times, and other clear indications of approval.

In Which is Revealed a Perverse Fact of my Personality

The sequel to my original lecture in 1979 reveals an aspect of my personality that my companions often find annoying. On 21 September 1979, Mary Beyda, assistant to the editor of the *American Psychologist*, Charles A. Kiesler, wrote saying among other things:

> Although he can't commit himself to publishing your address until he [Kiesler] has had a chance to see your manuscript, he is definitely interested in considering your work for publication in the February *American Psychologist*.

I sent two copies of my talk to Beyda with a covering letter dated 24 September 1979. A heavily edited copy of my manuscript was returned with an undated covering letter by Shelley M. Hammond. In writing my paper I had used the gramatically-correct language I had been taught and that I had always used up to that time, but it was, *Horrors!*, not *gender-free*. I returned my edited manuscript with my own marginal notes indicating in essence that, while I agreed with some of the changes made by Hammond, in her attempt to eliminate or circumvent sexist language she had made changes that were, in my opinion, awkward and worse than my original wording.

For example, she changed my "...concern with man's relationship to the total environment around him..." to "...concern with humans' relationship to the total environment around them..." My "Manpower requirements" was changed to "Human resources requirements." And so on.

This was followed by letters from Shelly Hammond to me, dated January 2, 1980, and a letter from Kiesler to me dated January 14, 1980. One sentence in Kiesler's letter, "I must insist on the change of the language", infuriated me. I withdrew my paper and in a letter dated 24 January 1980 said in part "You may think your expressions are good. But that's what you think. I don't like them. To me, they're awkward, contrived circumlocutions. Most important is that it's your style – not mine. Whose article is this, anyway?"

That did not end it, for I have a copy of a letter dated March 31, 1980 from Wilse Webb to Kiesler saying:

As you know I find much of the feminist activity simply ludicrous. Some I find to be an inappropriate imposition on the rights and beliefs of others. This, however, is an outrageous imposition on scientific communication.

And another from Charles N. Cofer to Kiesler on April 11, 1980, saying among other things:

I like the paper very much and would be sorry if it were to be lost to the *American Psychologist*. I think the APA guidelines on sexist language do not make the use of sexist language a condition requiring rejection of a Ms., and as a member of the task force I remember that the imposition of use of non-sexist language as one criterion for acceptance was rejected. ...Actually, the paper is not full of sexist language and such vestiges of it as the paper contains are referential, rather than evaluative.

Cofer also offered nine changes that he thought would satisfy both Kiesler and me.

Kiesler's final letter dated April 21, 1980, offered a compromise: If I would make the changes Cofer had suggested, he would accept the manuscript. At this point one facet of my personality, my intransigence, manifested itself. I never replied to Kiesler's letter.

That's the kind of uncompromising behavior I have exhibited on several other important occasions in my life, but I prefer not to write about them.

Another Oughta'

The reception to my presentation of this lecture at UConn told me that the paper, despite its age, is a good one and that it makes a number of points that are just as valid today as they were when I wrote it 14 years ago. What I really should do is change a few of my examples to more contemporary ones, computers, for example, and publish it. *Another* job for me!

HOW DO I SEE THEM NOW?

I enjoyed lecturing and I basked in the applause and plaudits that followed a successful lecture. But what good did all these lectures do for human factors? That's hard to say. I have always felt that public lectures are evanescent and, if they are not published and if listeners don't take notes, they and their messages are quickly forgotten. Those that I gave on my visiting professorships probably had some lasting impact because students *did* take notes. As for the rest, the most I could hope for is that they sensitized my listeners to human factors and what it can accomplish. I wonder, though, if in the long run it wouldn't have been better for me to have done less talking and more doing.

14. SOME OTHER PUBLICATIONS OF MORE THAN PASSING INTEREST 1958-1996

A number of my publications are not worth commenting on since they are either minor ones that had no survival value or publications with no special interest for me now. Here I describe some that I think are noteworthy.

A STUDY OF STOVE BURNERS

In 1958 I designed a rather simple experiment using a mockup of a stove. Four "burners" on the top surface were activated by four controls in a horizontal row across the front surface. The principal experimental variable was the linkage between controls and burners. I turned the apparatus over to Lionel Lindenbaum, an undergraduate student, who tested 60 subjects (15 on each of four linkages) for 80 consecutive trials. When I analyzed the data I found that two linkages turned out to yield faster response times and fewer errors than the worst two (Chapanis & Lindenbaum, 1959).

Ray and Ray (1979) performed the same kind of experiment with more arrangements than Lindenbaum and I had tested. It is gratifying that for those linkages matching ours their results agree with what we found. This is the kind of corroboration we seldom see in human factors. Unfortunately, there is too little incentive to repeat studies. For one thing editors don't like to publish them. Yet we can achieve true generality only by showing agreement among the results of studies performed in different laboratories, under diverse conditions, and with varied subjects (Chapanis, 1988b). So among other things, I argued in my 1988 article that

"If we are indeed seriously committed to increasing the store of true generalizations in our profession, we should encourage our students to replicate studies already done and should do so ourselves without feeling that we are wasting our time" (p. 266).

The Impact of Our Study

Although our stove study was a minor one it seems to have held some sort of special appeal because it has been cited over and over again and it has been the source of a number of studies with variations from the original (for example, Hsu & Peng, 1993; Osborne & Ellingstad, 1987; Payne, 1995; and Shinar & Acton, 1978). Our Figure 1 has also been reproduced in a number of textbooks (McCormick & Sanders, 1982, for example) and in December 1997, 38 years after the publication of our study, I received a request from the Galaxy Scientific Corporation for permission to reproduce the figure in their *Human Factors Guide for Aviation Maintenance,* a Federal Aviation Administration sponsored publication, released in 1998 in both hardcopy and on CD-ROM.

HOSPITAL MEDICATION ERRORS

In the late 1950s Charles D. Flagle from the Industrial Engineering Department directed a series of operational research studies in The Johns Hopkins Hospital. Impressed by accounts of errors made in the administration of medicines, he invited me to undertake a systematic study of the problem. I persuaded one of my students, Miriam A. Safren, to undertake this as a master's thesis. She and I designed a critical incident study with all the procedures needed for that purpose. She conducted all the briefings and collected the data. We analyzed the data and wrote an article together. When the work was completed, it was, I think, the first comprehensive and thorough analysis of errors of this kind (Safren & Chapanis, 1960).

Preparation for the Study

A great deal of preliminary work preceded the collection of data because one cannot simply ask nurses to report freely errors they have made. Instead, orientation sessions introduced the study to the supervisory personnel of all the services in the hospital. These were followed by a series of interviews with nurses in each division of the hospital. Perhaps the most important point Mrs. Safren emphasized was that all reports would be strictly anonymous and so could never be used for disciplinary or evaluation purposes.

Results

Over a period of seven months Mrs. Safren collected a total of 178 incidents, of which 143 (80%) were errors and 35 (20%) near errors. Some were frightening. For example, in some cases patients received or could receive ten times the amount of medication ordered, and in a few cases, 24 times the required dosage. The latter was due to a source of confusion that I have used time and again in lectures and in my writings. At that time prescriptions could be written with the symbols q.n. for the Latin *quaque nocte*, or "once every night." Other prescriptions might be written with the initials q.h. for the Latin *quaque hora*, or "once every hour." Whether a patient received the drug one time a day or 24 times a day depended on the correct reading of these easily confused handwritten symbols.

The three most frequent kinds of errors were (1) the wrong patient received or almost received a medication, (2) a patient received or almost received a wrong dose of medication, and (3) a patient received or almost received an extra (unordered) dose of medication. We analyzed the data according to their causes, job classification of the person committing the error or near error (for example, student nurse, graduate staff nurse, private duty nurse), time of day, and patient load. We concluded with a number of recommendations grouped under four major headings: (1) written communication, (2) medication procedures, (3) the working environment, and (4) training and education.

Significance of the Study

Although the critical incident technique has a number of

limitations I think this was a significant study that has been cited by others (for example, Fivars & Gosnell, 1966). It probably would have attracted more attention if it had been published in a journal more easily accessible to human factors personnel. Shortly after its publication, however, I was invited to prepare another article on this subject by the editor of the *Journal of Chronic Diseases* (Chapanis & Safren, 1960). It was also a source of satisfaction to me that several of our recommendations were actually implemented.

A Persistent Problem

The problem still persists, however, because newspapers from time to time still report accounts of such errors. Indeed, having been sensitized to the problem, I once stopped my own physician's nurse before she, due to a miscalculation, was about to inject me with ten times the intended dosage of a drug.

A PAPERBACK

In 1963 I was invited to write a paperback for inclusion in the Behavioral Science in Industry Series published by the Wadsworth Publishing Company. The 134-page book that I wrote (Chapanis, 1965a) barely scratched the surface of the field. Yet it was astonishingly successful and was published in the UK by Tavistock Publications and was translated and published in Spanish (Chapanis, 1968c), Japanese (Chapanis 1968d), Italian (Chapanis, 1970b), and Portuguese (Chapanis, 1972). Would that my latest book would sell that well!

ENCYCLOPEDIA ARTICLES

From 1968 to 1991 I was fortunate to have been invited to write 12 encyclopedia articles, eleven of them on human factors or engineering psychology (Chapanis, 1968a, 1970a, 1971a, 1973, 1974a, 1977a, 1977b, 1980a, 1980b, 1991a, 1991b) and one on color vision and color blindness (Chapanis, 1968b). I would like to believe that these articles had some influence in advancing the human factors profession.

A CRITICAL LOOK AT LABORATORY EXPERIMENTS

Most people entering the field of human factors in the 1950s and early 1960s came into it from psychology departments with an almost blind dedication to laboratory experimentation as the proper way to study human-machine relations. Although I was an experimentalist and have done many laboratory experiments, I felt and continue to feel that laboratory studies have many limitations when one tries to extrapolate from them to practical situations. So when I was invited to give a theme speech at the Third International Congress on Ergonomics I decided to write a sort of antidote, or at least a cautionary note, to the uncritical acceptance of laboratory studies as a way of solving practical problems.

The congress was organized by the Ergonomics Research Society on behalf of the International Ergonomics Association, and it was held at the University of Birmingham, England, from the 11th to the 15th of September, 1967. My speech (Chapanis, 1967) was received enthusiastically by some and with chagrin by others. In reading it today I find that it makes points that are still valid today. It was reprinted in a book by Schultz (1970a).

ENLARGING OUR HORIZONS

In the 1960s and 70s I felt that the practice of human factors in the US had become insular and narrowly focused on individual person-machine and systems problems with emphasis on efficiency and safety measured with performance criteria. In Europe, on the other hand, the practice of ergonomics struck me as being just as narrowly focused but on workers in heavy industry with emphasis on physiological indices of fatigue and energy expenditure. Convinced that human factors or ergonomic problems were much broader than were customarily being addressed, I tried to expand our horizons in three publications.

Cross-cultural Studies

In my lecturing and teaching during the late 40s and 50s I was sometimes asked, How general are these principles? Would these rules hold in all countries and cultures? My usual answer was that science and technology are universal and know no geographic boundaries. A considerable amount of foreign travel in the meantime led me to question that glib response and I became interested in organizing a symposium to explore what was known about human factors from a cross-cultural standpoint. Most granting agencies found the topic too applied to support, but I found a sympathetic ear in the Scientific Affairs Division of the North Atlantic Treaty Organization. After a formal proposal NATO agreed in 1971 to support a symposium on "National and Cultural Variables in Human Factors Engineering."

Figure 9. Participants at the *Symposium on National and Cultural Variables in Human Factors Engineering* held from the 19th to the 23rd of June 1972, in Oosterbeek, The Netherlands.

With the invaluable assistance of my co-organizer, Dr. John de Jong, of The Netherlands, the symposium was held from the 19th to the 23rd of June 1972, in the Hotel de Bilderberg[10] , Oosterbeek, The Netherlands. It was, all agreed, a huge success. Forty-four participants from fifteen countries attended. It was the first symposium ever organized on this topic, and it left no doubt that national and cultural variables are, or should be, significant factors in the application of human factors principles throughout the world. That symposium opened up what has since become an important area of research. I edited the papers presented at the symposium and published them in a book (Chapanis, 1975). A shorter paper (Chapanis, 1974b) describing some of the findings of

[10] We had taken over the entire hotel.

the symposium with some of my ideas about topics needing further research was presented as an invited address at the Fifth International Congress of Ergonomics, June 8, 1973, Amsterdam, The Netherlands.

Some Broader Societal Values

When I was invited to deliver a keynote address for the joint meeting of the 6th Congress of the International Ergonomics Association and the 20th Annual meeting of the Human Factors Society, I decided to call attention to some broader societal values. My address was given on July 11, 1976, in an auditorium at the University of Maryland, in College Park. Two coincidental events helped me to decide on my text. First, the year 1976 was the American Bicentennial Year. Second, we had had our first energy crisis just two years earlier.

I began my talk by describing briefly the adoption of the Declaration of Independence on July 4, 1776, and quoted the following from that document:

> We hold these truths to be self-evident, that all men are created equal, that they are endowed by their Creator with certain unalienable Rights, that among these are Life, Liberty, and the pursuit of Happiness.

I used that quotation as the theme of my talk, that we had to look beyond the traditional, rather narrow focus of simple human-machine systems to much broader societal values and concerns: energy conservation, the population explosion, environmental pollution and the quality of life. I believe my talk was the first to call attention to the relevance of these broader societal issues to ergonomics and to the ways ergonomics might help in dealing with them.

Reactions to my speech. I always thought that my introduction was a neat way of tying my talk to the Bicentennial Year and to the theme of what I wanted to convey. Unexpectedly it elicited adverse reactions for two different reasons. Several colleagues from our mother country, England, found my reference to the Declaration of Independence offensive. On the other hand, Jan Rosner, a good friend and delegate from Poland, still then under Russian dominance, asked if I would give him permission to translate and republish my talk in Poland, but without referring to the Declaration of Independence and especially to the quotation about life, liberty and the pursuit of happiness. I might add, however, that in making that request I feel sure Rosner (now deceased) was only trying to be politically correct. After all, although it was not well known, he had been a member of the Polish underground and had fought first against Germany and then against Russia in World War II.

Social Consequences of Ergonomics

I began my keynote address to the Seventh Congress of the International Ergonomics Association, held in Warsaw, Poland, August 27, 1979, with a salutation to a great Polish novelist, Henryk Sienkiewicz (Chapanis, 1979a). From the title of one of his novels, for which he received a Nobel price in 1905, I adapted his words, *Quo vadis, domine?* Where are you going, Master? for the title and theme of my talk, *Quo vadis, ergonomia?* Where are you going, ergonomics?

I answered that question by saying that ergonomics is beginning to become involved in questions of social planning and social strategy that have far-reaching consequences. Some of the decisions we make affect not only our own society, but other societies as well, and the consequences of decisions made now may extend beyond our individual lifetimes. I used examples from two quite different areas of application – safety and office automation – to illustrate the wide ranging social consequences of ergonomic decisions. I argued that the involvement of ergonomics in these broad questions is healthy, but that it brings with it greatly increased responsibilities.

I ended my talk by saying that the answer to my initial question was also to be found in the same work by Sienkiewicz. My question was *Quo vadis, ergonomia?* Where are you going, ergonomics? and the succinct answer is *Urbi et orbi*, to the city and to the world.

MY CREDO

An invitation to give the keynote address to the Human Factors Association of Canada/Association Canadienne d'Ergonomie at its annual meeting in Edmonton, Canada, on September 14, 1988, provided me with what seemed like the perfect opportunity to deal with recent concerns within our profession about our failure to communicate our findings to designers, the inadequacy of our data for solving practical design problems, and the lack of generality of most of our findings (Chapanis, 1990). In reading that speech now, I find it a clear statement of my philosophy about human factors, human factors research, and the way we should communicate our findings to designers. It may be the best speech I ever wrote.

212

Defining Human Factors

For several years before I gave my speech, a number of us who were primarily practitioners had been concerned about the kinds of articles that were being published in *Human Factors*. They were often basic research articles that had little or no relevance to human factors. In another related article written shortly after my Canadian talk (Chapanis, 1991e), I gave some examples from the publication called *PsycSCAN: APPLIED EXPERIMENTAL & ENGINEERING PSYCHOLOGY*, a collection of abstracts published periodically by the American Psychological Association. The following are titles under the heading "Human Factors & Ergonomics" in one issue of *PsychSCAN*:

1. Optical and photoreceptor immaturities limit the spatial and chromatic vision of human neonates.
2. "Pure alexia" without hemianopia or color anomia.
3. Detection of visual stimuli after lesions of the superior colliculus in the rat; deficit not confined to the far periphery.
4. Is obesity an eating disorder?

I called these a hodgepodge of miscellaneous and irrelevant articles. I made clear that I did not intend to criticize the contents of any of these articles. What I deplored was their inclusion in the category of human factors. It indicated to me that many people did not understand what human factors was, a problem that had concerned me for years (Chapanis, 1969b). That led to my definition of the field:

Human Factors (Ergonomics) is a body of knowledge about human abilities, human limitations, and other human characteristics that are relevant to design.

What we do is human factors engineering, which I defined in this way:

Human Factors Engineering (The practice of ergonomics) is the application of human factors (ergonomic) information to the design of tools, machines, systems, tasks, jobs, and environments for safe, comfortable and effective human use.

I have never been satisfied with most other definitions of human factors and human factors engineering. Some are catchy but so short that they really don't define the field and what we do. Others are, to my mind, verbose, tortured, and complex. Although it may be a sign of vanity I have always liked mine because they are short, clear, and say it all. Incidentally, as is apparent, I have always considered human factors and ergonomics to be synonymous.

The Critical Word: Design

I pointed out that the critical word in both definitions is *design*, because it is this that separates us from such purely academic disciplines as psychology, physiology, and anthropology. I argued that research, even so-called basic research, qualifies as human factors research only if it is oriented toward the design of something. In both my 1990 and 1991e articles I gave examples from the *Human Factors* journal of articles that did not contribute to the solution of any design problem. I stated, and feel strongly, that if the author(s) cannot find any design implications in an article, that article does not belong in the *Human Factors* journal.

At the same time, I described briefly several articles that were technically sound and relevant to human factors, but from which the author(s) did not draw out any design implications. We cannot, I said, expect engineers or designers to read our minds and deduce the design implications of what we have done. If there are design implications in what we do, it is our responsibility to say what they are. For that reason, every article published in our journal should contain a final section pointing out the design implications of the study.

The State of Human Factors Knowledge

I next addressed criticisms that there has been little or no accumulation of general human factors knowledge. On this matter I was more optimistic than many critics have been. I called attention to the hundreds of human factors guidelines that had already been published and stated that there are generalizations behind all these guidelines, even though they might not be apparent from the way guidelines are usually stated.

I next pointed out that many of our generalizations, recommendations and guidelines (for example, "The red light used for external lighting shall have a luminous intensity of 2.5 candela, an X chromaticity coordinate not less than 0.650, and a Y chromaticity coordinate not greater than 0.330.") are precise enough for design. Designers know how to meet those requirements because they are precisely stated. Moreover, an inspector can easily verify whether a machine or system has components that meet those requirements.

On the other hand, many guidelines (for example, "The choice of the type of dialogue between the user and the computer...shall be compatible with user characteristics and task requirements.") are not precise enough for design because they leave it to the designer to decide, in this case, what the user characteristics and task requirements are and what kind of dialog is compatible with them. It is our responsibility as human factors professionals to translate guidelines of that kind into precise project-specific requirements that engineers and designers can use without any further interpretation.

On the Need to do Studies

The last major point I tried to make concerns the frequently heard complaint that human factors always wants to do another study. That is not a major fault of our field because I didn't think we could ever come with general rules that will apply to every conceivable kind of application and the diverse users for whom each application might be designed. As an example, I said that I didn't think we could conceive of a generalized vocabulary that would be appropriate for automated bank tellers, washing machines, power lawn mowers, automobiles, nuclear power plants, aircraft, and spacecraft. The applications are too diverse and the potential users vary too much. Not only are applications diverse but they change as technology advances. The answer, I said, is that we are always going to have to do some kind of study to come up with specific design recommendations.

Finally, I said that I don't think we need to be apologetic about having to do additional studies of one kind or another to come up with specific design recommendations. Engineers do it all the time. No engineer goes directly from an idea or design concept into full-scale production of a finished product. Engineers constantly prepare models, simulations, breadboards, and prototypes to test, modify, and validate their own design decisions. In doing our own brand of research we are doing no more than is accepted in the practice of good design engineering.

Evaluation

I still feel strongly about all the things I said in my speech, because I think they help to clarify and define who we are, what we do, and how we do it. It has apparently had some impact because I find that editors for *Human Factors* are beginning to look for design implications in articles that are submitted to them. My speech was received with much approbation in Canada. I was also pleased when it was reprinted in its entirety in the *Human Factors Society Bulletin* (Chapanis, 1991 & 1992).

HUMAN FACTORS DEFICIENCIES IN HOSPITALS

The nine-year period from 1987 to 1996 was a time during which I suffered grievous assaults on both my psyche and soma. Six hip replacements[11], nine hip dislocations, replacement of my ascending aorta with a Dacron tube, and two other minor operations kept me

[11] No, I don't have six legs. One hip replacement had to be surgically corrected four times.

in and out of four different hospitals and one rehabilitation center for extended periods of time.

During seemingly endless confinements and despite intense pain and heavy sedation I repeatedly found examples of poor human factors design that could have been anticipated or corrected with some common sense principles. One simple example will suffice.

> After hip surgery patients are warned against bending forward or down. Yet in all four hospitals paper dispensers in the toilets were placed in awkward positions, usually so low that paper could be retrieved only by bending down or by calling a nurse or aide for assistance.

I occupied myself by writing notes about these design deficiencies and later prepared a short article (Chapanis, 1996a) that was published in *Ergonomics in Design*. I received more sympathetic compliments, some from as far away as New Zealand, about that small article than I ever received after the publication of many more substantial articles. I had obviously put my finger on problems that clearly need human factors attention. I concluded by saying that "no one should ever be allowed to design a hospital until he or she has undergone major surgery and has spent ten days or more in a hospital bed" (p. 36).

A RECAPITULATION

I'm satisfied with this potpourri of articles. Each was good in its own way and every one has attracted enough attention[12] for me to feel secure in saying that they have contributed in at least a small way to the advancement of our profession. That's a good feeling.

---

[12] Just as I was about to send this manuscript off to my publisher, the July 1998 issue of Ergonomics in Design arrived in the mail. In it I found an article by Andre and Segal and on page 5 an illustration and discussion of the misplaced toilet paper holder problem about which I had written. Since they do not cite my 1996 article, I cannot, of course, claim that they were influenced by me. I can at least claim that they agree with me that this is a serious design fault for individuals who are handicapped either because of medical problems or advanced age.

15. A FEW ODDS AND ENDS
1963-1993

This chapter describes a few activities that did not seem to fit into earlier ones.

A STUDY OF THE ACCOUNTING PROFESSION

One of the men for whom I had great respect and admiration was professor Robert H. Roy, Dean of the School of Engineering Science at Johns Hopkins. In 1963 Roy and James H. MacNeill, Chairman of the Department of Accounting, School of Business Administration, Fordham University, were asked to conduct a study of the accounting profession and, in particular, to identify "the common body of knowledge for certified public accountants" for the coming dynamic future.

Their first attempt to identify and rank the relative

importance of various subject matters by means of a "Subject Matter Evaluation Manual" was unsuccessful (Roy & MacNeill, 1967, pp. 173-176). At that point they asked for my help (p. 176). I suggested a forced-choice procedure: put each of the 53 subject matter content areas on a single card, shuffle the cards, and ask each of a number of respondents to arrange the cards in order of importance. This was termed the card deck experiment and it turned out to yield a great deal of important and usable data. After the data had been collected, I provided advice about the way they should be presented and analyzed.

Although this was not a major achievement and had nothing to do with human factors, it gave me a great deal of satisfaction because (1) my suggestions had worked so well and (2) I had been able to help a man for whom I had a great deal of respect. I was also pleased by the acknowledgement of my contribution in the text of the book and by this gracious note in the forward: "In the design of the card deck experiment, we had the sagacious counsel of Dr. Alphonse Chapanis, Professor of Psychology, The Johns Hopkins University."

THE NATIONAL RESEARCH COUNCIL
COMMITTEE ON HUMAN FACTORS

The Committee on Human Factors was established in October 1980 and I was honored to have been a member of the committee from 1980 to 1985.

Applied Methods in Human Factors

Early in my tenure I suggested that we needed (1) to

systematize the methods used in applied human factors work, (2) more books and reference sources on these methods, and (3) better education about these methods. As a starting point I chaired a workshop on applied methods in December 1981. Participants were Stuart K. Card, Xerox Palo Alto Research Center; David Meister, US Navy Personnel Research and Development Center; Donald L. Parks, Boeing Aerospace Company; Richard W. Pew, Bolt, Beranek & Newman, Inc.; Erich P. Prien, Memphis State University; John B. Shafer, IBM Corporation; and Robert T. Hennessy, National Research Council. Hennessy and I wrote a report on that workshop (Chapanis & Hennessy, 1985) and the workshop motivated Meister to write his excellent book on the subject (Meister, 1985).

Other Contributions

I also helped prepare a chapter on user-computer interaction (Chapanis, Anderson & Licklider, 1985) and on population group differences (Goldstein & Chapanis, 1985). The background of the latter article is an example of the kind of thinking that I was constantly battling in the committee. Goldstein has been asked to write a chapter on population group differences and he did just that. I criticized it severely because at no point had he drawn out the consequences of these differences for human factors. What does it mean for human factors that there are large differences between the sexes, or between ethnic groups? I then illustrated what I meant by pointing out various examples in which equipment had to be designed differently to accommodate the sexes, people of diverse nationalities, and other groups that differed from one another. In the end, Goldstein invited me to co-author the chapter with him since he didn't feel that he could draw out the human factors implications as well as I could.

I Become a Gadfly

I was constantly reminding the committee that we were a committee on *human factors* and that basic research was not basic research in human factors unless we could show that it had *design implications*. Several of the academic members of the committee could not seem to understand the distinction I was trying to make and I remember some rather heated arguments with some of them about this matter. I don't think I overstate the case when I say that my constant needling helped to keep the committee focused and on track.

THE BOARD OF CERTIFICATION
IN PROFESSIONAL ERGONOMICS

For a long time I felt strongly that we needed qualification standards for the profession and some mechanism for certifying those individuals who met those standards, a point of view I expressed most clearly in my keynote address to the 29th Annual meeting of the Human Factors Society (Chapanis, 1985). It is not surprising then that Dieter Jahns should invite me to be a member of a board attempting to set up an organization for testing and certifying human factors professionals. Since my file of activities has been incinerated, I can no longer write about any details of what I did.

I believe, however, that I was instrumental in framing the definition of human factors and ergonomics that is used by the board. I also know that I was busy for years, writing memos and letters and attending meetings. Among other things, I had long debates with David Meister about the nature of the examinations we should have to test competency. Dave wanted a kind of essay examination in which candidates would be required to write how they would solve a particular problem and in so doing demonstrate the ability to analyze, design, and test. He was also opposed to any kind of objective examination questions. I argued that to be certified candidates should have knowledge of a basic core of subject matter material, for example, statistics, and that the best way to do that was with objective examination questions. Since I have never seen a recent examination, I don't know what they consist of these days, but I believe that objective questions do form at least part of the examinations.

The Name

Jahns had initially wanted to call this organization an Academy. I argued that it should be called a Board, which designation was finally adopted.

Human Factors or Ergonomics?

Jahns had also wanted to certify individuals only as *ergonomists.* I argued strongly for certifying individuals either as human factors professionals or ergonomists, with the choice of designation to be made by the person being certified. This idea was finally adopted and a person may now choose to be a *CPE, Certified Professional Ergonomist,* or a *CHFP, Certified Human Factors Professional.*

Two Testimonials

I was immensely pleased when the Board honored me by granting me a certificate as BCPE Certificant Number 1 on 30 September 1992. In the spring of 1994 while I was at home convalescing from one of my several operations Hal Hendrick telephoned to ask if he could visit. When he did it was to present me with a handsome plaque dated 10 December 1993 and addressed to "Alphonse Chapanis, Ph.D., CHFP, Co-founder and Director Emeritus, in Appreciation for services provided to establish professional practice policies & procedures." I cannot put into words how touched I was.

A FINAL WORD

Of the three activities described here, the formation of the Board of Certification in Professional Ergonomics (BCPE) is by far the most important. Through its certification process and examinations, the Board has helped to define and clearly establish human factors as a discipline with a body of knowledge that is unique to it and with a purpose that is distinct from that of both engineering and the academic disciplines on which it draws. I also believe that the Board has increased the professionalism of practitioners in human factors and gained new respect for our profession from persons in other disciplines. I am glad and proud to be able to say that I played a small part in the formation of the BCPE.

16. A CONFESSION, AN EPISODE, AND AN IRONICAL AFTERMATH

Since I am now at the end of my story and near the end of my life I would like to acknowledge for the first time publicly that some of my trips to foreign countries were financed by an agency of the US intelligence community for which I wrote classified reports about certain individuals, organizations, and conditions I observed or encountered during my travels.

AN EPISODE

One of my trips into the USSR is perhaps worth a few words because it was marked by several incidents which, although vexing at the time, seem almost hilarious today. On this trip my wife (now ex-wife) was an unwitting ally because she is of Russian heritage and speaks Russian fluently.

All trips at the time had to be arranged through Intourist, the official Russian tourist agency. I planned visits to Leningrad (now

St. Petersburg), Moscow, and Yerevan. Yerevan, the capital of Armenia, is about 1,100 miles south of Moscow and may appear to be an unusual place for tourists to visit. As it turns out, however, there were a number of interesting tourist sites there. More important for this account, however, is that it was the center of one of the most important Russian institutes of mathematics and automation. For our trip I paid in advance for first class accommodations and for a car, chauffeur, and translator-guide in all three cities.

Our Entry into the USSR

Since I had to stop off in London and Paris before proceeding to the USSR, my wife and I traveled separately and met in Helsinki. We stayed there overnight and took a plane the next day into Leningrad. Arrival at the airport was hectic. Russia had not yet entered the electronic age, at least insofar as travel was concerned. The immigration officials had long lists of passengers' names unalphabetized and written out in long hand which they scanned laboriously as each person presented his or her passport. When it came my turn, the official could not find my name! I insisted that she keep searching while I watched from across the counter. And then I found it! I had been taking some Russian classes at the University in anticipation of this trip and I was able to recognize my first name written in Russian script. On the arrival list my name had been written first name first and the official had been searching for my last name.

I Discover An Easy Way to Call for Room Service

After several hours those of us who had been duly cleared to enter the country took a bus into the city and were deposited at our respective hotels, in our case the Ukraina. In my briefing before leaving the US I had been told that we would undoubtedly be assigned Room 701 and that that room was bugged. After unpacking, I went into the bathroom to wash up and found skimpy bath towels, more like what we would call kitchen towels at home. This provided me an opportunity to test the system. I came out of the bathroom and said loudly to my wife something like "Look at these towels. How do they expect us to dry ourselves? Is this the best this country can give first-class tourists?" About ten minutes later, there was a loud knock on our door. I opened the door to find a large lady holding an arm load of real bath towels!

You Just Have to Know How to Ask

When we were ready to do some sightseeing, we went down to the Intourist desk in the lobby of the hotel and found it crowded with tourists all demanding service of one kind or another in a half dozen different languages. When I was finally able to approach the desk, I presented my travel documents and asked for our car, chauffeur, and translator. "No car. Take bus." was the reply I received. I remonstrated that I had paid for a private car, chauffeur, and translator. The official was unmoved. "No car. Take bus!" After one more interchange with the same result, my wife took over.

In clear and loud Russian she told the official that we were important visitors, that we were being treated shabbily, and that we would go back to America and tell everyone that travel agents in Russia were dishonest. As my wife went on and on indignantly in this vein, the official grew red and finally interrupted, saying in

228

Russian, "Take seat, please." About 15 minutes later we had a car, chauffeur, and translator! And we had these services every day of our visit.

Our travels were interesting because my wife usually did not at first reveal that she spoke Russian. We would sit in the back seat of the car and she would whisper to me what the driver and translator were saying to each other. In this way I objected to some places they intended to take us and was able to direct them to places I wanted to go, sometimes by insisting. On some occasions, though, they never could seem to find an address I was interested in. I wondered if they ever suspected the truth about us.

Well Dressed Maintenance Men

In Moscow we had an experience that showed we were under surveillance. We had left our room, but in the lobby I remembered that I had forgotten something. I returned to our room just in time to see a man leaving it. He murmured something like "repair man" in Russian and hurried on. Maintenance man, indeed! To believe that I would have to believe that maintenance men in Russia all wore suits.

A Plea for Asylum

I had introductions to a number of people in Moscow. On one occasion, a prominent scientist we were visiting took me into his study, turned up the volume on his radio, and said in German, the only common language we had, that he wanted to defect to America. I replied that we were only tourists and that I had nothing to do with that kind of activity. I said I would take his name and pass it on to someone in the US who might be interested. I found out much later that he had indeed emigrated to the US.

We Enjoy Singular Limousine Service

Food at that time was scarce and the variety limited. I hungered for fresh fruit the whole time we were in the USSR. Identical menus with extensive offerings were handed to us in every restaurant we went to, but we quickly learned that the restaurant was out of almost everything on the menu. Ordering was actually a matter of selecting from the few items a waiter identified as being available. Service was agonizingly slow. Since waiters were all state employees and could not be fired they had little incentive to accommodate patrons.

One evening one of our hosts offered to take us to a special restaurant for dinner. The special restaurant turned out to be an underground restaurant (almost literally) serving black market food. Our hosts had one of the taxis in front of our hotel drive us to what was labeled a private soccer club in the basement of a large building. When we were identified and admitted by a doorkeeper, we found a busy dining room and enjoyed the best meal we had during our entire visit.

Several hours later we emerged from the "club," climbed a few stairs to street level to find it almost entirely deserted. Vehicles of almost every variety were, of course, very scarce. Standing on the curb I agonized about how we could get back to our hotel when our host flagged down a cruising ambulance. He exchanged some words with the driver and then turned to me saying, "Geev heem five rubles." A short time later we were deposited on the street in front of our hotel. Ambulance drivers evidently earned extra income moonlighting in this way.

Yerevan

Arriving in Yerevan, we had a repetition of no car, no chauffeur, no translator, resolved again with my wife's eloquence. With another introduction that had been arranged in advance, I was able to visit the institute I wanted to see, but without my wife. About 10 staff members and I sat around a table and exchanged information through a translator. Although I'm sure I did not get the same information I would have were my wife present, I did learn some interesting things. My sponsor was interested in the names of some of the individuals engaged in research there – who does what – and I was successful in that respect. I was also able to see one of their computer systems which they claimed was used for controlling their space exploration flights. I found it interesting that the system was programmed in English and used English, rather than Russian commands. I also thought this system was primitive by American standards at that time.

I Narrowly Escape the Gulag

We had arrived in Yerevan after dark. Since our room was not air conditioned, we slept with the windows wide open. In the morning, I could see that our room faced a large square in the center of which there was a large statue of Stalin on a raised platform. I decided to do a little exploring before breakfast, went out to the square, mounted the platform to put my hand on the statue to try to decipher the inscription. Suddenly I was roughly seized by a policeman. I frantically waved my passport in his face shouting

"Ya Amerikanski tourist." The altercation brought my wife to my rescue again. He was finally persuaded that I must be just an ignorant but innocent foreigner and let me go with a stern warning to stay away from the statue.

I Prepare for Our Return

By the time we returned to Moscow I had a rather large collection of handwritten notes that I did not want to have among our belongings when we left, so I took them to the American embassy and turned them over to the military attaché. He, in turn, had them sent back to the US in a diplomatic pouch.

AN IRONICAL AFTERMATH

Paradoxically, years later these activities caused me to lose my SECRET clearance and a potentially interesting consulting job because I could not explain satisfactorily to another agency why I had taken so many trips abroad. So much for the rewards of patriotism. Or, perhaps I should say, this shows that one arm of the government didn't know what another arm had done.

Nonetheless, if my unnamed sponsors should happen to read this, I would like them to know that my assignments added a lot of spice and a little bit of risk to what would otherwise have been some routine scientific trips. I thank them for having made my life even more exciting than it normally was.

17. EPILOGUE

Having extracted everything I want to leave on record from those of my chronicles I can remember, how do I feel about it? In reading this manuscript all the way through in one sitting my immediate reaction is one of utter astonishment verging on incredulity. Did I really do all the things I say I did? To reassure myself that I have not made it all up, however, I can turn to the incontrovertible evidence provided by articles, newspaper accounts, stamps in my passports, and other records still in my possession. I *really* did everything I say I did. But just reading it all today exhausts me.

In looking back on what I have written here I see things that I regret having said and done and things that I would do differently if I had the chance to relive my life. I am also chagrined to see research data, carefully preserved through three retrenchments, that I have never found the time to analyze fully, write up, and publish. Then there are articles started and never finished. Now that I have time, I move so much more slowly, have so few resources and so little energy that I will probably never complete them. For that I am truly sorry.

But there is one thing I have never regretted – and that is my choice of profession. Human factors has always been challenging, frustrating at times, rewarding at others, but never dull. I can honestly say in retrospect that I have had a full life – an exciting life – and that I have enjoyed telling people about human factors, educating students and others to take over where I have had to leave off, and grappling with the problems of trying to make our material world safer, more comfortable, and easier to cope with. In fact, there is only one thing I truly regret –

I'm sorry I've come to the end.

REFERENCES

These references[13] do not include everything I've written. A number of my articles and reports are not important enough to merit being listed here.

Al-Awar, J., Chapanis, A., & Ford, W. R. (1981). Tutorials for the first-time computer user. *Institute of Electrical and Electronics Engineers Transactions on Professional Communication*, PC-24, 30-35.

Al-Awar, J. A. (1985). *An Instrument for Evaluating the Use of Speech in Computer Communication*. Doctoral dissertation. Baltimore, MD: The Johns Hopkins University.

ANSI/HFS 100-1988. *American National Standard for Human Factors Engineering of Visual Display Terminal Workstations*. Santa Monica, CA: The Human Factors Society, Inc.

Bernotat, R., & Hunt, D. P. (Eds.) (1977). *University Curricula in Ergonomics*. Meckenheim, Germany: Forschungsinstitut für Anthropotechnik.

Brecht, M. A. (1979). *Study of Meeting and Conference Behavior*. Doctoral dissertation. Baltimore, MD: The Johns Hopkins University.

Chapanis, A. (1937). A note on the validity and difficulty of items in Form A of the Otis Self-Administering Tests of Mental Ability. *Journal of Experimental Education*, 5, 246-248.

Chapanis, A. (November 4, 1942). *Luminescent Materials*.

---

[13] Because of the age and variety of many of the unpublished documents among these references, I cite titles, dates, and sources in full and exactly as they appear on the documents. For them I have not followed conventional standard forms of citation.

Memorandum Report Number EXP-M-49-695-12F. Dayton, OH: Army Air Forces Materiel Center, Wright Field.

Chapanis, A. (25 November 1943). *Optical Distortion in the B-25 Bombardier's Sighting Window*. Memorandum Report No. ENG-49-695-39. Dayton, OH: Army Air Forces Materiel Center, Aero medical Laboratory, Wright Field.

Chapanis, A. (1944a). Spectral saturation and its relation to color vision defects. *Journal of Experimental Psychology*, 34, 24-44.

Chapanis, A. (7 May 1944). *Optical Distortion in Transparent Sections of the B-29*. Memorandum Report No. ENG-49-695-37E. Dayton, OH: Army Air Forces Materiel Center, Engineering Division, Aero Medical Laboratory, Wright Field.

Chapanis, A. (12 May 1944). *The Quantitative Measurement of Visual Fields*. Technical Report No. 5112. Dayton, OH: Army Air Forces, Materiel Center, Wright Field.

Chapanis, A. (1945). Night vision – a review of general principles. *The Air Surgeon's Bulletin*, 2, 279-284.

[Chapanis, A.] (15 March 1945). Vision. Chapter VIII, pp. 59-67, in *Physiology of Flight: Human Factors in the Operation of Military Aircraft*. AAF Manual No. 25-2. Washington, D.C.: Headquarters, Army Air Forces.

Chapanis, A. (1 July 1945). *Night Vision Testing and Training in the Army Ground Forces*, Memorandum Report No. TSEAL3-695-48E. Dayton, OH: Army Air Forces Air Technical Service Command, Engineering Division, Aero Medical Laboratory, Wright Field.

Chapanis, A. (1946a). A device for demonstrating the effects of anoxia on vision. *Journal of Aviation Medicine*, 17, 348-356.

Chapanis, A. (1 November 1946). *Speed of Reading Target Information from a Direct-Reading Counter-Type Indicator*

References

Versus Conventional Radar Bearing-and-Range Dials. Memorandum Report No. 166-I-3. Jamestown, RI: Systems Research Field Laboratory.

Chapanis, A. (1946b). The dark adaptation of the color anomalous. *American Journal of Physiology*, 146, 689-701.

Chapanis, A. (1 August 1947). *The Relative Efficiency of a Bearing Counter and Bearing Dial for Use with PPI Presentations.* Memorandum Report No. 166-I-26. Jamestown, RI: Systems Research Field Laboratory.

Chapanis, A. (1947). The dark adaptation of the color anomalous measured with lights of different hues. *Journal of General Physiology*, 30, 423-437.

Chapanis, A. (1948a). A comparative study of five tests of color vision. *Journal of the Optical Society of America*, 38, 626-649.

Chapanis, A. (1948b). An attempt to construct a quantitative pseudo-isochromatic test of color vision. (Abstract). *The American Psychologist*, 3, 245.

Chapanis, A. (20 June 1949). *Some Aspects of Operator Performance on the VJ Remote Radar Indicator*. Memorandum Report No. SDC 166-I-91. Baltimore, MD: Psychological Laboratory, Institute for Cooperative Research, The Johns Hopkins University.

Chapanis, A. (1949a). Diagnosing types of color deficiency by means of psudo-isochromatic tests. *Journal of the Optical Society of America*, 39, 242-249.

Chapanis, A. (1949b). The stability of "improvement" in color vision due to training – a report of three cases. *American Journal of Optometry and Archives of American Academy of Optometry*, 26, 251-259.

Chapanis, A. (1949c). Simultaneous chromatic contrast in normal and abnormal color vision. *American Journal of Psychology*, 62, 526-539.

Chapanis, A. (1950). Relationships between age, visual acuity and color vision. *Human Biology*, 22, 1-33.

Chapanis, A. (May 27, 1950). An evaluation of the Navy color vision lantern. Pp. 109-140 in *Minutes and Proceedings of the Subcommittee on Color Vision, Twenty-Sixth Meeting of the Armed Forces – NRC Vision Committee*. Ottawa, Canada.

Chapanis, A. (1951a). Color blindness. *Scientific American*, 184, 48-53.

Chapanis, A. (1951b). Theory and methods for analyzing errors in man-machine systems. *Annals of the New York Academy of Sciences*, 51, 1179-1203.

Chapanis, A. (November 16, 1951). An experimental determination of some iso-color lines in color-deficient vision. Pp. 24-36 in *Minutes and Proceedings of the Twenty-Ninth Meeting of the Armed Forces – NRC Vision Committee*. New London, CT: US Navy Submarine Base.

Chapanis, A. (January 1953). *An Evaluation of Problems of Chart Reading Under Red Illumination*. Ann Arbor, MI: University of Michigan.

Chapanis, A. (1956). *The Design and Conduct of Human Engineering Studies*. San Diego, CA: San Diego State College Foundation.

Chapanis, A. (1959). *Research Techniques in Human Engineering*. Baltimore, MD: The Johns Hopkins Press.

Chapanis, A. (1961). Men, machines, and models. *American Psychologist*, 16, 113-131.

Chapanis, A. (1963a). Engineering psychology. Pp. 285-318 in P. R. Farnsworth, O. McNemar, & Q. McNemar (Eds.), *Annual*

Review of Psychology, Volume 14. Palo Alto, CA: Annual Reviews.

Chapanis, A. (1963b). From the President. *Human Factors Society Bulletin*, 6(10), 1.

Chapanis, A. (1963c). From the President. *Human Factors Society Bulletin*, 6(12), 1.

Chapanis, A. (1964a). From the President... At last, a home of our own! *Human Factors Society Bulletin*, 7(3), 1.

Chapanis, A. (1964b). From the president. *Human Factors Society Bulletin*, 7(9), 1.

Chapanis, A. (1965a). *Man-Machine Engineering*. Belmont, CAP: Wadsworth Publishing Co.

Chapanis, A. (1965b). Words, words, words. *Human Factors*, 7, 1-17.

Chapanis, A. (1965c). On the allocation of functions between men and machines. *Occupational Psychology*, 39, 1-11.

Chapanis, A. (1967). The relevance of laboratory studies to practical situations. *Ergonomics*, 10, 557-577.

Chapanis, A. (1968a). Engineering psychology. Pp. 81-87 in D. L. Sills (Ed.), *International Encyclopedia of the Social Sciences*, Volume 5. New York, NY Macmillan.

Chapanis, A. (1968b). Color vision and color blindness. Pp. 329-336 in D. L. Sills (Ed.) *International Encyclopedia of the Social Sciences*, Volume 16. New York, NY: Macmillan.

Chapanis, A. (1968c). *Ingenieria Hombre-Maquiná*. Mexico, D.F.: Compañia Editorial Continental, S.A.

Chapanis, A. (1968d). *Ningen to Kikai*. Tokyo: Diamond.

Chapanis, A. (1969a). Symposium on criteria in man-machine systems. *Ergonomics*, 12, 951-952.

Chapanis, A. (1969b). Human factors – the rudderless ship. *Human Factors Society Bulletin*, 12(7), 1.

Chapanis, A. (1970a). Human engineering. Pp. 545-547 in *The Encyclopedia Americana; International Edition*, Volume 14. New York, NY: Americana Corporation.

Chapanis, A. (1970b). *L'Ergonomia: Introduzione allo Studio dei Sistemi Uomo-Macchina*. Milan, Italy: Franco Angeli Editore.

Chapanis, A. (1970c). Relevance of physiological and psychological criteria to man-machine systems. *Ergonomics*, 13, 337-346.

Chapanis, A. (1971a). Human engineering. Pp. 677-680 in *Encyclopedia of Occupational Health and Safety*, Volume 1, A-K. Geneva, Switzerland: International Labour Office.

Chapanis, A. (1971b). The search for relevance in applied research. Pp. 1-14, in W. T. Singleton, J. G. Fox, & D. Whitfield (Eds.) *Measurement of Man at Work*. London: Taylor and Francis.

Chapanis, A. (1972). *A Engenharia e o Relacionamento Homen-Máquina*. Sao Paulo: Brasil: Editora Atlas, S.A.

Chapanis, A. (1973). Engineering psychology. Pp. 85-88 in *Psychology '73/'74 Encyclopedia*. Guilford, CT: Dushkin Publishing Group.

Chapanis, A. (1974a). Human factors engineering. Pp. 1168-1169 in *Encyclopedia Britannica*, Volume 8. Chicago, IL: Macropaedia, Encyclopedia Britannica.

Chapanis, A. (1974b). National and cultural variables in ergonomics. *Ergonomics*, 17, 153-175.

Chapanis, A. (Ed.) (1975). *Ethnic Variables in Human Factors Engineering*. Baltimore, MD: Johns Hopkins Press.

Chapanis, A. (1976a). *Evaluating the Costs of Alternative Office Systems*. Glen Arm, MD: Alphonse Chapanis, Ph.D., P.A.

Chapanis, A. (1976b). Ergonomics in a world of new values. *Ergonomics*, 19, 253-268.

Chapanis, A. (1977a). Human engineering. Pp. 545-547 in *Encyclopedia Americana*, Volume 14. New York, NY: American Corporation.

Chapanis, A. (1977b). Engineering psychology: an overview. Pp. 330-332 in *International Encyclopedia of Psychiatry, Psychology, Psychoanalysis, and Neurology*, Volume IV. Birmingham, AL: Aesculapius Publishers.

Chapanis, A. (1979a). Quo Vadis, Ergonomia. *Ergonomics*, 22, 595-605. Also *Ergonomia*, (In Polish) 2, 109-122.

Chapanis, A. (1979b). Interactive communication: a few research answers for a technological explosion. Pp. 33-67 in M. Amirchahy & D. Neel (Eds.) *Nouvelles Tendances de la Communication Homme-Machine*. Le Chesnay, France: Institut National de Recherche en Informatique et en Automatique.

Chapanis, A. (1980a). Mensch-Maschine-Systeme in Produktion und Verkehr. Pp. 736-759 in F. C. Stoll (Ed.) *Die Psychologie des 20. Jahrhunderts, Volume XIII: Psychologie im Berufsleben*. Zürich, Switzerland: Kindler Verlag.

Chapanis, A. (1980b). Mensch-Maschine-Systeme im Alltag. Pp. 760-778 in F. C. Stoll (Ed.) *Die Psychologie des 20. Jahrhunderts. Volume XIII: Psychologie im Berufsleben*. Zürich, Switzerland: Kindler Verlag.

Chapanis, A. (1980c). *Hours of Work, Work Schedules, and Fatigue in Certain Chesapeake Bay Piloting Operations*. Glen Arm,

MD: Alphonse Chapanis, Ph.D., P.A.

Chapanis, A. (1981). Interactive human communication: some lessons learned from laboratory experiments. Pp. 65-114, in B. Shackel (Ed.) *Man-Computer Interaction: Human Factors Aspects of Computers & People*. Alphen aan den Rijn, The Netherlands: Sijthoff & Noordhoff.

Chapanis, A. (1982a). Computers and the common man. Pp. 106-132, in R. A. Kasschau, R. Lachman, & K. R. Laughery (Eds.) *Information Technology and Psychology: Prospects for the Future*. Praeger.

Chapanis, A. (1982b). Man-computer research at Johns Hopkins. Pp. 238-249 in R. A. Kasschau, R. Lachman, & K. R. Laughery (Eds.) *Information Technology and Psychology: Prospects for the Future*. Praeger.

Chapanis, A. (1982c). Humanizing computers. Pp. 21-53 in *Proceedings: Human Factors Symposium, 18th-19th May 1982*. Essex, UK: ITT Europe Engineering Support Centre.

Chapanis, A. (1984). Taming and civilizing computers. *Annals of the New York Academy of Sciences*, 426, 202-218.

Chapanis, A. (1985). Some reflections on progress. Pp. 1-8 in *Proceedings of the 29th Annual Meeting of the Human Factors Society*. Santa Monica, CA: Human Factors Society.

Chapanis, A. (1986). A psychology for our technological society: or, a tale of two laboratories. Chapter 3, pp. 53-70 in S. H. Hulse & Bert F. Green, Jr. (Eds.) *One Hundred Years of Psychological Research*. Baltimore, MD; The Johns Hopkins Press.

Chapanis, A. (1988a). Should you believe what the new ANSI/HFS standard says about numeric keypad layouts? No! *Human Factors Society Bulletin*, 31(11), 6-9.

Chapanis, A. (1988b). Some generalizations about generalization.

Human Factors, 30, 253-267.

Chapanis, A. (1990). To communicate the human factors message you have to know what the message is and how to communicate it. *Communiqué*, 21(2), 1-5, and 21(3), 1-4. Reprinted in *Human Factors Society Bulletin*, 1991, 34(11), 1-4 and 1992, 35(1), 3-5.

Chapanis, A. (1991a). Human factors in computer systems. Pp. 305-324 in A. Kent and J. G. Williams (Eds.) *Encyclopedia of Microcomputers. Volume 8: Geographic Information System to Hypertext*. New York, NY: Marcel Dekker.

Chapanis, A. (1991b). Human factors in computer systems Pp. 150-170 in A. Kent (Ed.). *Encyclopedia of Library and Information Science*. Volume 48. New York, NY: Marcel Dekker.

Chapanis, A. (1991c). The business case for human factors in informatics. Chap. 3, pp. 39-71, in B. Shackel & S. J. Richardson (Eds.) *Human Factors for Informatics Usability*. Cambridge, UK: Cambridge University Press.

Chapanis, A. (1991d). Evaluating usability. Chap. 16, pp. 359-395, in B. Shackel & S. J. Richardson (Eds.) *Human Factors for Informatics Usability*. Cambridge, UK: Cambridge University Press.

Chapanis, A. (1991e). Making human factors truly human factors. *CSERIAC Gateway*. 2(3), 1-3.

Chapanis, A. (1995). Human production of "random" numbers. *Perceptual and Motor Skills*, 51, 1347-1363.

Chapanis, A. (1996a). Musings from a hospital bed. *Ergonomics in Design*, 4, 35.

Chapanis, A. (1996b). *Human Factors in Engineering Design*. New York, NY: Wiley.

Chapanis, A., Anderson, N. S., & Licklider, J. C. R. (1983). User-computer interaction. Chapter 5, pp. 78-124, in Committee on Human Factors, Commission on Behavioral and Social Sciences and Education, National Research Council: *Research Needs for Human Factors*. Washington, D. C.: National Academy Press.

Chapanis, A., & Budurka, W. J. (1990). Specifying human-computer interface requirements. *Behaviour and Information Technology*, 9, 479-492.

Chapanis, A., Garner, W. R., Morgan, C. T., & Sanford, F. H. (1947). *Lectures on Men and Machines: An Introduction to Human Engineering*. Baltimore, MD: Systems Research Laboratory.

Chapanis, A., Garner, W. R., & Morgan, C. T. (1949). *Applied Experimental Psychology: Human Factors in Engineering Design*. New York, NY: Wiley.

Chapanis, A., & Gropper, B. A. (1968). The effect of the operator's handedness on some directional stereotypes in control-display relationships. *Human Factors*, 10, 303-319.

Chapanis, A., & Halsey, R. M. (1953). Photopic thresholds for red light in an unselected sample of color-deficient individuals. *Journal of the Optical Society of America*, 43, 62-63.

Chapanis, A., Hartline, H. K., Larrabee, M. G., & De Lucchi, M. R. (March 1953). *A study of Visual Reconnaissance*. Baltimore, MD: Human Factors Scientific Panel, The Johns Hopkins University.

Chapanis, A., & Hennessy, R. T. (1983). Applied methods in human factors. Chapter 7, pp. 140-160, in Committee on Human Factors, Commission on Behavioral and Social Sciences and Education, National Research Council: *Research Needs for Human Factors*. Washington, D. C.: National Academy Press.

Chapanis, A., & Lindenbaum, L. L. (1959). A reaction time study of four control-display linkages. *Human Factors*, 1, 1-7.

References

Chapanis, A., Marsh, B. W., & Viteles, M. S. (June 14, 1950). *Conclusions and Recommendations of the National Research Council Conference on Highway Safety Research.* Unpublished memorandum submitted to the National Research Council.

Chapanis, A., Rouse, R. O., & Schachter, S. (1949). The effect of inter-sensory stimulation on dark adaptation and night vision. *Journal of Experimental Psychology, 39,* 425-437.

Chapanis, A., & Safren, M. A. (1960). Of misses and medicines. *Journal of Chronic Diseases, 12,* 403-408.

Chapanis, A., & Shafer, J. B. (1986). Factoring humans into FSD systems. *Technical Directions, 12*(1), 15-22.

Chapanis, N. P., & Chapanis, A. (1964). Cognitive dissonance: five years later. *Psychological Bulletin, 61,* 1-22.

Conrad, R., & Hull, A. J. (1968). The preferred layout for numeral data-entry keysets. *Ergonomics, 11,* 165-173.

Deininger, R. L. (1960). Human factors engineering studies of the design and use of pushbutton telephone sets. *Bell System Technical Journal, 39,* 995-1012.

Dvorine, I. (1944a). A new diagnostic method of testing and training color perception. *American Journal of Optometry and Archives of American Academy of Optometry, 21,* 225-235.

Dvorine, I. (1944b). Reconditioning the color-blind – a case report. *American Journal of Optometry and Archives of American Academy of Optometry, 21,* 508-510.

Dvorine, I. (1946). Improvement in color vision in twenty cases. *American Journal of Optometry and Archives of American Academy of Optometry, 23,* 302-321.

Fitts, P. M. , & Jones, R. E. (July 1, 1947). *Analysis of Factors Contributing to 460 "Pilot Error" Experiences in Operating*

Aircraft Controls. Memorandum Report No. TSEAA-694-12. Wright-Patterson Air Base, OH: Army Air Forces Air Materiel Command, Engineering Division, Aero Medical Laboratory.

Fitts, P. M (Ed.), Chapanis, A., Frick, F. C., Garner, W. R., Gebhard, J. W., Grether, W. F., Henneman, R. H., Kappauf, W. E., Newman, E. B., & Williams, A. C. Jr., (March, 1951). *Human Engineering for an Effective Air-Navigation and Traffic-Control System*. Washington, D. C.: National Research Council, Division of Anthropology and Psychology, Committee on Aviation Psychology.

Fivars, G., & Gosnell, D. (1966). *Nursing Education: The Problem and the Process: The Critical Incident Technique*. New York, NY: Macmillan.

Fleishman, E. A. (Ed.) (1967). *Studies in Personnel and Industrial Psychology*. Homewood, IL: Dorsey Press.

Gagge, A. P. (January 9, 1943). *Emergency Flares*. Inter-Office Memorandum AC:zb:49. Dayton, OH: Army Air forces, Materiel Command, Wright Field.

Goldstein, I. L., & Chapanis, A. (1983). Population group differences. Chapter 6, pp. 125-139 in Committee on Human Factors, Commission on Behavioral and Social Sciences and Education, National Research Council: *Research Needs for Human Factors*. Washington, D.C.: National Academy Press.

Green, R. J., Self, H. C., & Ellifritt, T. S. (Eds.) (1965). *50 Years of Human Engineering: History and Cumulative Bibliography of the Fitts Human Engineering Division*. Wright-Patterson Air Force Base, OH: Crew Systems Directorate, Armstrong Laboratory, Air Force Materiel Command.

Halsey, R. M., & Chapanis, A. (1952). An experimental determination of some iso-chromaticity lines in color-deficient vision. *Journal of the Optical Society of America*, 42, 722-739.

Hanson, B. J. (1983). A brief history of applied behavioral science at Bell Laboratories. *The Bell System Technical Journal, 62,* 1571-1590.

Hardy, L. H., Rand, G., & Rittler, M. C. (1957). *AO H-R-R Pseudoisochromatic Plates for Detecting, Classifying, and Estimating the Degree of Defective Color Vision.* Southbridge, MA: American Optical-Instrument Division.

Hartley, C. S., Brecht, M. A., Pagerey, P. D., Weeks, G. D., Chapanis, A., & Hoecker, D. G. (1977). Subjective time estimates of work tasks by office workers. *Journal of Occupational Psychology, 50,* 23-36.

Hsu, S-H., & Peng, Y. (1993). Control/display relationship of the four-burner stove: a reexamination. *Human Factors, 35,* 745-749.

Human Factors Scientific Panel (March 1953). *A Study of Visual Reconnaissance.* Baltimore, MD: The Johns Hopkins University.

Johns Hopkins University Circular (March, 1941). *The College of Arts and Sciences.* Catalog Number 1941-1942, Whole Number 523, New Series, 1941, Number 3. Baltimore, MD: University.

Johns Hopkins University Circular (March, 1942). *The College of Arts and Sciences.* Catalog Number 1942-1943, Whole Number 532, New Series, 1942, Number 3. Baltimore, MD: University.

Johns Hopkins University Circular (March, 1943). *The College of Arts and Sciences.* Catalog Number 1943-1944, Whole Number 541, New Series, 1943, Number 3. Baltimore, MD: University.

Kekceev[14], K. Ch. (1937). On the action of non-adequate stimuli on receptors. *Comptes-rendus de l'Academie des Sciences de Russie, 14,* 495-497.

Kekcheev, K. K. (1943). Methods of accelerating dark adaptation

---

[14] This author's name has been transliterated in several ways. The spellings I use here are the ones that appear in the original articles.

and improving night vision. *War Medicine*, 8, 209-220.

Kektcheeff, K. C., & Astrovsky, E. P. (1941). Sur la définition des vibrations aériennes de fréquence ultra-sonore par les mesures des seuils optiques. *Comptes-rendus de l'Academie des Sciences de Russie*, 31, 370-372.

Kelley, J. F. (1983). *Natural Language and Computers: Six Empirical steps for Writing an Easy-to-use Computer Application*. Doctoral dissertation. Baltimore, MD: The Johns Hopkins University.

Klemmer, E. T., & Haig, K. A. (1988). Weight and balance of a new telephone handset. *Applied Ergonomics*, 19, 271-274.

Lasareff, P. P. (1937). Théorie thermique des changements de la sensibilité visuelle périphérique produits par les causes géophysique. *Comptes-rendus de l'Academie des Sciences de Russie*, 14, 275-278.

Lasareff, P. P., & Dobrovolskaia, E. V. (1937). De l'influence du chant sur l'adaptation au cours de la vision périphérique. *Comtes-rendus de l'Academie des Sciences de Russie*, 14, 271-272.

Lutz, M. C., & Chapanis, A. (1955). Expected locations of digits and letters on ten-button keysets. *Journal of Applied Psychology*, 39, 314-317.

Marx, M. H. (Ed.) (1963). *Theories in Contemporary Psychology*. New York, NY: Macmillan.

McCollom, I. V., & Chapanis, A. (1956). *A Human Engineering Bibliography*. San Diego, CA: San Diego State College Foundation.

McCormick, E., J., & Sanders, M. S. (1982). *Human Factors in Engineering and Design*. (Fifth Ed.). New York, NY: McGraw-Hill.

References

Meister, D. (1985). *Behavioral Analysis and Measurement Methods*. New York, NY: Wiley.

MIL-STD-1472E (31 October 1996). *Department of Defense Design Criteria Standard*. Washington, D.C.: Department of Defense.

Morgan, C. T., Cook, J. S., III, Chapanis, A., & Lund, M. W. (Eds.) (1963). *Human Engineering Guide to Equipment Design*. New York, NY: McGraw-Hill.

ORI (4 March 1974). *A Program of Research on Personnel Qualifications, Training and Licensing for the Improvement of Merchant Marine Safety*. Technical Memorandum 113-74. Silver Spring, MD: Author.

Osborne, D. W., & Ellingstad, V. S. (1987). Using sensor lines to show control-display linkages on a four-burner stove. Pp. 581-584 in *Proceedings of the Human Factors Society 31st Annual Meeting*. Santa Monica, CA: Human Factors Society.

Panel on Psychology and Physiology, Committee on Undersea Warfare (1949). *A Survey Report on Human Factors in Undersea Warfare*. Washington, D.C.: National Research Council.

Parsons, J. M (1972). *Man-Machine System Experiments*. Baltimore, MD: Johns Hopkins Press.

Payne, S. J. (1995). Naive judgments of stimulus-response compatibility. *Human Factors, 37*, 495-506.

Pinson, E. A., & Chapanis, A. (1945). AML portable radium plaque night vision tester. *The Air Surgeon's Bulletin, 2*, 285.

Pinson, E. A., & Chapanis, A. (1946). Visual factors in the design of military aircraft. *Journal of Aviation Medicine, 17*, 115-122.

Pinson, E. A., Romejko, W. J., & Chapanis, A. (1945). Flying sun glasses with rose smoke lens. *The Air Surgeon's Bulletin, 2*, 141.

Potosnak, K. M. (1983). *Choice of Computer Interface Modes by Empirically Derived Categories of Users*. Doctoral dissertation. Baltimore, MD: The Johns Hopkins University.

Ray, R. D., & Ray, W. D. (1979). An analysis of domestic cooker control design. *Ergonomics*, 22, 1243-1248.

Roemer, J. M (1981). *Learning Performance and Attitudes as a Function of the Reading Grade Level of a Computer-Presented Tutorial*. Doctoral dissertation. Baltimore, MD: The Johns Hopkins University.

Roscoe, S. N. (1997). *The Adolescence of Engineering Psychology*. Santa Monica, CA: Human Factors and Ergonomics Society.

Roy, R. R., & MacNeill, J. H. (1967). *Horizons for a Profession*. New York, NY: American Institute of Certified Public Accountants.

Safren, M. A., & Chapanis, A. (1960). A critical incident study of hospital medication errors. *Hospitals*, 34(9), 32-34 et passim and 34(10), 53 et passim.

Scales, E. M., & Chapanis, A. (1954). The effect on performance of tilting the toll-operator's keyset. *Journal of Applied Psychology*, 38, 452-456.

Schmidt, R. A., & Lee, T. (Eds.) (1998). *Motor Control and Learning*. (3rd Ed.) Champaign, IL: Human Kinetics.

Schultz, D. P. (Ed.) (1970a). *Psychology and Industry*. New York, NY: Macmillan.

Schultz, D. P. (Ed.) (1970b). *The Science of Psychology: Critical Reflections*. New York, NY: Appleton-Century-Crofts.

Shinar, D., & Acton, M. B. (1978). Control-display relationships on the four-burner range: population stereotypes versus standards. *Human Factors*, 20, 13-17.

Sinaiko, H. W. (Ed.) (1951). *Selected Papers on Human Factors in the Design and Use of Control Systems*. New York, NY: Dover Publications.

Singleton, W. T., Fox J. G., & Whitfield, D. (Eds.) (1971). *Measurement of Man at Work: An Appraisal of Physiological and Psychological Criteria in Man-Machine Systems*. London, UK: Taylor and Francis.

Sleight, R. B. (1948). The effect of instrument dial shape on legibility. *Journal of Applied Psychology*, 32, 170-188.

Southwell, E. A., & Merbaum, M. (Eds.) (1964). *Personality: Readings in Theory and Research*. Belmont, CA: Wadsworth Publishing Company.

Streltsov, V. (1944). The function of the eye in aviation. *American Review of Soviet Medicine*, 2, 127-133.

Systems Research Laboratory (November 1, 1945). *Motion and Time Analysis of Combat Information Centers: Tucson-Nashville-Louisville-Boston*. Research Report Number 4. Cambridge, MA: Harvard University.

Turugin, S. J. (1937). Effect of electromagnetic centimetre waves on the central nervous system. *Comptes-rendus de l'Academie des Sciences de Russie*, 17, 19-21.

Tyler, M. (1982). New life for voice mail (Editorial). *Datamation*, 28(13), 53-55.

Van Cott, H. P., & Kinkade, R. G. (Eds.) (1972). *Human Engineering Guide to Equipment Design*. (Revised Edition). Washington, DC: U.S. Government Printing Office.

Venturino, M. (Ed.) (1990). *Selected Readings in Human Factors*. Santa Monica, CA: Human Factors Society.

Weeks, G. D., Hartley, C. S., Chapanis, A., Brecht, M. A., Pagerey, P., & Hoecker, D. G. (1975). *Studying Work Activities in Office Systems*. Glen Arm, MD: Alphonse Chapanis, Ph.D., P.A.

Wexley, K. N., & Yukl, G. A. (Eds.) (1975). *Organizational Behavior and Industrial Psychology*. New York, NY: Oxford University Press.

Wogalter, M. S., Begley, P. B., Scancorelli, L. F., & Brelsford, J. W. (1997). Effectiveness of elevator service signs: measurement of perceived understandability, willingness to comply and behavior. *Applied Ergonomics*, 28, 181-187.

Woodson, W. E. (1954). *Human Engineering Guide for Equipment Designers*. Berkeley, CA: University of California Press.

Yuki, G. A., & Wexley, K. N. (Eds.) (1971). *Readings in Organizational and Industrial Psychology*. New York, NY: Oxford University Press.

Zoltan, E., & Chapanis, A. (1982). What do professional persons think about computers? *Behavior and Information Technology*. 1, 55-68.

APPENDIX:
CERTIFICATES AND OTHER
AWARDS I HAVE RECEIVED

February 8, 1943
Diploma from Yale University: "Praeses et socii Universitatis Yalensis in novo portu in re publica Connecticutensi Omnibus ad quos hae litterae pervenerint salutem in domino sempiternam nos praeses et socii huius Universitatis Alphonse Chapanis amplioris honoris academici candidatum ad gradum titulumque Philosophiae Doctoris admisimus eique concessimus omnia iura privilegia insignia ad hunc honorem spectantia in cuius rei testimonium his litteris universitatis sigillo impressis nos praeses et scriba academicus subscripsimus A.D. VIII Id. Feb. MDCCCCXXXXIII

1960
"The Society of Engineering Psychologists a Division of the American Psychological Association Certificate of Appreciation to Alphonse Chapanis President 1960 in recognition of outstanding contribution to the science and profession of engineering psychology"

September 1961
Office of Naval Research, Washington, D.C. "This Certificate of Appreciation is award to Alphonse Chapanis in recognition of his contributions to the advancement of U.S. Naval Research and international scientific cooperation while serving as a civilian scientific officer in London from June 1960 to September 1961"

1963
Plaque from the Society of Engineering Psychologists inscribed "Alphonse Chapanis The Franklin V. Taylor Award for Outstanding Contributions to the field of Engineering Psychology"

October 1973
Plaque inscribed "The Human Factors Society presents The Paul M. Fitts Award to Alphonse Chapanis for outstanding contributions to human factors education"

Undated (1976)
"The Ergonomics Honorary Fellowship Awarded to Alphonse Chapanis."

1976
Certificate stating "Be it known that Alphonse Chapanis is a Fellow of the Human Factors Society an international organization dedicated to increasing

253

knowledge about man and his performance in relation to equipment and working environments and to the application of human factors knowledge to the design of systems, equipment and devices"

October 1977

Plaque inscribed "The Human Factors Society presents The President's Distinguished Service Award to Alphonse Chapanis for a prestigious career involving dedicated service and leadership to the human factor profession"

1978

"American Psychological Association Distinguished Contribution for Application in Psychology Award 1978 presented to Alphonse Chapanis for his contributions as a founder of the field of engineering psychology and for his pioneering research and leadership in the field over a 35 year period. He was the senior author (with Wendell Garner and Clifford Morgan) of the first systematic book to cover the field of engineering psychology and he wrote the first important methodology text in the field. The enormous range of his contributions include his early studies on basic visual mechanism, his research on work-station design and man-machine systems relations, and his more recent studies on information processing and telecommunications. He has provided numerous insights on ways to apply sound psychological research to societal problems in a technological age."

May 1981

Plaque inscribed "Maryland Psychological Association Outstanding Scientific Contributions to Psychology Alphonse Chapanis, Ph.D."

1981

"The Human Factors Society presents this Certificate of Appreciation to Alphonse Chapanis In Special Recognition of his Long, Faithful and Outstanding Service: Member of Executive council 1961-1981, President 1963-1964"

1982

IEEE Professional Communication Society Outstanding Article Award "Tutorials for the First-Time Computer User by Janan Al Awar, Alphonse Chapanis, W. Randolph Ford. The authors have contributed significantly to professional communication by identifying a serious communication need, meeting that need with a workable plan, and communicating the plan with precision and enthusiasm"

1982

Plaque from the International Ergonomics Association inscribed "In recognition of Outstanding Contributions to Ergonomics Internationally"

1982

Wojciech Bogumit Jastrzebowski 1799-1882 Precursor Ergonomiae Medal "To Professor Alphonse Chapanis from the Polish Ergonomics Society in appreciation of merits for international ergonomics"

February 22, 1991

Framed Certificate of Recognition "Presented to Alphonse Chapanis for Outstanding Achievement, and Contribution to the IBM Owego Business as